T0259697

NIMS Monographs

National Institute for
Materials Science

NIMS publishes specialized books in English covering from principle, theory and all recent application examples as NIMS Monographs series. NIMS places a unity of one study theme as a specialized book which was specialized in each particular field, and we try for publishing them as a series with the characteristic (production, application) of NIMS. Authors of the series are limited to NIMS researchers. Our world is made up of various "substances" and in these "materials" the basis of our everyday lives can be found. Materials fall into two major categories such as organic/polymeric materials and inorganic materials, the latter in turn being divided into metals and ceramics. From the Stone Ages - by way of the Industrial Revolution - up to today, the advance in materials has contributed to the development of humankind and now it is being focused upon as offering a solution for global problems. NIMS specializes in carrying out research concerning these materials. NIMS: http://www.nims.go.jp/eng/index.html

More information about this series at http://www.springer.com/series/11599

Takahiro Nagata

Nanoscale Redox Reaction at Metal/Oxide Interface

A Case Study on Schottky Contact and ReRAM

 Springer

Takahiro Nagata
Research Center for Functional Materials
National Institute for Materials Science
Tsukuba, Ibaraki, Japan

ISSN 2197-8891 ISSN 2197-9502 (electronic)
NIMS Monographs
ISBN 978-4-431-54849-2 ISBN 978-4-431-54850-8 (eBook)
https://doi.org/10.1007/978-4-431-54850-8

This Springer imprint is published by the registered company Springer Japan KK part of Springer Nature.
The registered company address is: Shiroyama Trust Tower, 4-3-1 Toranomon, Minato-ku, Tokyo
105-6005, Japan

Preface

Today, integrated circuit technologies are at a crucial turning point in the establishment of the fundamentals for further advancement. To overcome the performance limits of conventional materials such as the SiO_2 gate, polycrystalline Si gate, and Al wires, it is necessary to develop a new material with a new functionality not yet found in Si devices. Oxide materials are a good candidate to replace Si devices since these materials show exotic properties in accordance with the composition design and/or doping technique. These materials should realize future functional devices with high-k, ferroelectric, magnetic, and optical properties.

In this book, we discuss the investigation and intentional control of the metal/oxide interface structure and electrical properties using the data obtained by nondestructive methods such as X-ray photoelectron spectroscopy (XPS). At the metal/oxide interface, oxygen plays an important role in redox reactions, and thus affects the electrical properties. To observe and control these properties, a metal Schottky contact for an optical sensor application (Chaps. 1 and 2) and a metal/oxide resistive random-access memory structure (Chaps. 3 and 4) are investigated. For example, nitrogen plasma treatment on an oxide surface can reduce the surface electron accumulation layer, resulting in an enhancement of the Schottky property of a metal/oxide interface. Additionally, in Chap. 5, we briefly introduce combinatorial thin-film synthesis, which is our specialty and a suitable method for exploring new materials. While we do not attempt to cover every single aspect of oxide research in this book, we do aim to present discussions on selected topics that are both representative and possibly of technological interest.

We expect this book to be of interest to scientists and engineers working in the field of metal oxides.

Tsukuba, Japan   Takahiro Nagata

Acknowledgments

First of all, I would like to acknowledge the editorial teams of Springer and NIMS Monographs for their patience and valuable suggestions. I am deeply grateful to my many colleagues at NIMS, the BL15XU NIMS Beamline, and Meiji University for invaluable suggestions, discussions, and technical support.

I am also grateful to HiSOR, Hiroshima University, and JAEA/SPring-8 for developing HX-PES at BL15XU, SPring-8. The HX-PES measurements were performed under the approval of the NIMS Beamline Station (BL-15XU) (Proposal nos. 2007B4604, 2009A4600, 2009B4601, 2010A4604, 2010B4600, 2011A4611, 2011B4611, and 2012A4613). Part of the work in this book was supported by Kakenhi Grant-in-Aid for Scientific Research B 19760224, a Grant-in-Aid for Key Technology, "Atomic Switch Programmed Device", and the World Premier International Research Center Initiative (WPI), from Japan's Ministry of Education, Culture, Sports, Science, and Technology (MEXT) Japan.

Contents

Chapter 1
General Introduction

The integrated circuit technologies stand at a crucial turning point in establishing the fundamentals for further advancement. To overcome the performance limits of conventional materials such as SiO_2 gate, polycrystalline Si gate, and Al wiring, it is necessary to develop a new material with a new functionality that is not found in Si devices up to now, such as nonvolatile memory function. Various material combinations and device structures have been proposed and investigated regardless of in-organic or organic materials. In this field, our interest is in in-organic compound materials, especially oxides. From the point of view of the electrical device application, oxide ceramic and crystals are widely used in bulk form for more than half a century, such as ceramics capacitor, electro-optical switch, surface acoustic wave devices, and so on. In the film case, the practical application in electrical devices was limited in the use of a transparent conducting oxide (TCO) due to its transparency in the visible range and relatively high electrical conductivity except for SiO_2 [1, 2]. In the late 1990s, hafnium oxide-based high-k oxide and $InGaZnO_x$ (IGZO) accelerated the thin-film oxide application in the electrical devices [3, 4]. High-k oxide replaces the SiO_2 gate material and enhances the scale down rule limitation, with the so-called Moore technology. In contrast, IGZO archived the lower power consumption in the liquid crystal display application due to its transparency and mobility instead of the Si transistor. These applications focused spotlight on other oxide materials. In the case of semiconductor, wideband gap oxides such as SnO_2, ZnO, and In_2O_3 exhibit a highly sensitive surface which has proven useful as a sensor technology [5–7]. Recently, additional various electrical applications such as electrode materials in displays [8], light-emitting diodes (LEDs) [9], and transparent thin-film transistors (TTFTs) [10, 11] have also been investigated.

From the point of view of the multifunctional oxide materials, recently resistive random-access memory (ReRAM) has been proposed as a new application for oxide materials due to its simple structure. An oxide sandwiched between two metal electrodes shows reversible electric field-induced resistance switching behaviors. For nonvolatile memory applications, ferroelectric materials have been researched for two decades. However, typical ferroelectric materials indicated the scale effect, and some of the ferroelectric materials, including the hazardous or alkali metals such

© National Institute for Materials Science, Japan 2020
T. Nagata, *Nanoscale Redox Reaction at Metal/Oxide Interface*, NIMS Monographs,
https://doi.org/10.1007/978-4-431-54850-8_1

as Pb, Li, and Ca, are not suitable for the Si-based film electronics. In contrast, oxide-based ReRAM structure indicated material flexibility and scale down ability. Several types of ReRAM structure have been demonstrated. A typical resistive switching model is based on a thermal effect initiated by a voltage-induced partial dielectric breakdown that forms a discharge filament modified by Joule heating [12, 13]. The intrinsic material properties also induce changes in resistance. For example, the insulator–metal transition in perovskite oxides such as $(Pr,Ca)MnO_3$ [14–16] and $SrTiO_3$:Cr [17] is induced by electronic charge injection operations such as doping. These materials indicate the scale down ability that contributes to the high-density device integration.

For these thin-film nanoelectronics device applications, with decreasing device scale, the importance of interface structure is pronounced, which is strongly related to the formation of the oxygen vacancies. At the oxide semiconductor surface, oxygen vacancies induced the Fermi-level pinning, which can be controlled by post deposition treatments such as plasma treatment. At the metal/oxide interface, oxygen plays an important role in redox reactions, affecting the electrical properties. To observe and control them, metal Schottky contact for an optical sensor application and metal/oxide resistive random-access memory structure are investigated. In this book, the investigation and intentional control of metal/oxide interface structure and electrical properties with the data obtained by nondestructive methods such as X-ray photoelectron spectroscopy (XPS) are discussed; for example, nitrogen plasma treatment on an oxide surface. These treatments can reduce the surface electron accumulation layer, resulting in an enhancement of Schottky property of a metal/oxide interface. This book consists of five chapters based on our research results as follows:

Chapters 2 and 3: The interface structure of Schottky metal on oxide semiconductor.
As an example, we chose zinc oxide (ZnO) as an oxide semiconductor since ZnO is a polar material. The metal and oxygen terminated surface can be obtained, which is suitable for the investigation of the oxygen effects on the interface.
Chapters 4 and 5: The oxygen effect on the switching properties of ReRAM structure.
As an example, we chose hafnium oxide (HfO_2)-based ReRAM structure. HfO_2 is used as a high-k gate insulator for advanced complementary metal/oxide semiconductor (CMOS) technologies; it has shown resistance switching phenomena, and there has been of increased interest in the use of HfO_2 and related oxides as potential ReRAM materials.
Chapter 6: Brief introduction of combinatorial thin-film synthesis
In the former chapters, the combinatorial thin-film synthesis plays an important role for systematic and high-throughput analysis. To help readers understand the sample fabrication and analysis, the combinatorial thin-film synthesis is introduced briefly.

While the book does not attempt to cover every single aspect of oxides research, it does aim to present discussions on selected topics that are both representative and possibly of technological interest.

References

1. Minami T (2005) Transparent conducting oxide semiconductors for transparent electrodes. Semicond Sci Technol 20:S35. https://doi.org/10.1088/0268-1242/20/4/004
2. Ginley DS, Bright C (2000) Transparent conducting oxides. MRS Bull 25:15. https://doi.org/10.1557/mrs2000.256
3. Hosono H, Kikuchi N, Ueda N, Kawazoe H (1996) Working hypothesis to explore novel wide band gap electrically conducting amorphous oxides and examples. J Non-Cryst Solids 198–200:165. https://doi.org/10.1016/0022-3093(96)80019-6
4. Nomura K, Ohta H, Ueda K, Kamiya T, Hirano M, Hosono H (2003) Thin-film transistor fabricated in single-crystalline transparent oxide semiconductor. Science 300:1269. https://doi.org/10.1126/science.1083212
5. Kohnke EE (1962) Electrical and optical properties of natural stannic oxide crystals. J Phys Chem Solids 23:1557. https://doi.org/10.1016/0022-3697(62)90236-6
6. Nagasawa M, Shionoya S, Makishim S (1965) Vapor reaction growth of SnO_2 single crystals and their properties. Jpn J Appl Phys 4:195. https://doi.org/10.1143/JJAP.4.195
7. Choudhary J, Ogale SB, Shinde SR, Kulkarni VN, Vendatesan T, Harshavardhan KS, Strikovski M, Hannoyer B (2004) Pulsed-electron-beam deposition of transparent conducting SnO_2 films and study of their properties. Appl Phys Lett 84:1483. https://doi.org/10.1063/1.1651326
8. Batzill M, Katsiev K, Burst JM, Diebold U, Chaka AM, Delley B (2005) Gas-phase-dependent properties of SnO_2 (110), (100), and (101) single-crystal surfaces: structure, composition, and electronic properties. Phys Rev B 72:165414. https://doi.org/10.1103/PhysRevB.72.165414
9. Anisimov OV, Gaman VI, Maksimova NK, Mazalov SM, Chernikov EV (2006) Electrical and gas-sensitive properties of a resistive thin-film sensor based on tin oxide. Semiconductors 40:704. https://doi.org/10.1134/S1063782606060170
10. Kim H, Pique A, Horwitz JS, Mattoussi H, Murata H, Kafafi ZH, Chrisey DB (1999) Indium tin oxide thin films for organic light-emitting devices. Appl Phys Lett 74:3444. https://doi.org/10.1063/1.124122
11. von Wenckstern H, Splith D, Lanzinger S, Schmidt F, Müller S, Schlupp P, Karsthof R, Grundmann M (2015) pn-hetero diodes with n-type In_2O_3. Adv Electr Mater 1·1400026. https://doi.org/10.1002/aelm.201400026
12. Pagnia H, Sotnik N (1988) Bistable switching in electroformed metal–insulator–metal devices. Phys Stat Sol (a) 108:11. https://doi.org/10.1002/pssa.2211080102
13. Chudnovskii FA, Odynets LL, Pergament AL, Stefanovich GB (1996) Electroforming and switching in oxides of transition metals: the role of metal–insulator transition in the switching mechanism. J Solid State Chem 122:95. https://doi.org/10.1006/jssc.1996.0087
14. Asamitsu A, Tomioka Y, Kuwahara H, Tokura Y (1997) Current switching of resistive states in magnetoresistive manganites. Nature 388:50. https://doi.org/10.1038/40363
15. Fors R, Khartsev SI, Grishin AM (2005) Giant resistance switching in metal-insulator-manganite junctions: evidence for Mott transition. Phys Rev B 71:045305. https://doi.org/10.1103/PhysRevB.71.045305
16. Kim DS, Kim YH, Lee CE, Kim YT (2006) Colossal electroresistance mechanism in a $Au/Pr_{0.7}Ca_{0.3}MnO_3/Pt$ sandwich structure: evidence for a Mott transition. Phys Rev B 74:174430. https://doi.org/10.1103/physrevb.74.174430
17. Meijer GI, Staub U, Janousch M, Johnson SL, Delley B, Neisius T (2005) Valence states of Cr and the insulator-to-metal transition in Cr-doped $SrTiO_3$. Phys Rev B 72:155102. https://doi.org/10.1103/PhysRevB.72.155102

Chapter 2
Changes in Schottky Barrier Height Behavior of Pt–Ru Alloy Contacts on Single-Crystal ZnO

2.1 Introduction

Zinc oxide (ZnO), a wide band-gap II–VI semiconductor, has major potential for use in optical devices. For example, it has attracted much attention for potential use in light-emitting and light-detecting devices in the ultraviolet (UV) region [1–3]. Although similar devices based on gallium nitride (GaN) have been put to practical use, ZnO has a number of advantages over GaN, including higher quantum efficiency, high exciton binding energy at room temperature, greater resistance to high-energy radiation, and amenability to wet chemical etching [4]. High-quality and thermally reliable Schottky contacts are crucial. We have proposed a UV-region Schottky-type photodiode using ZnO [5].

Compared with the formation of an ohmic contact on ZnO, the formation of a Schottky contact on ZnO is complicated, owing to the very high donor concentration in the surface region, which consists of native defects such as oxygen vacancies and zinc interstitials [6]. Allen et al. prepared a high-quality AgO Schottky contact on ZnO by using a conventional simple surface cleaned by organic solvents [7]. Au and Pt Schottky contacts on bulk ZnO wafers have shown good Schottky properties [8], but the electrical properties of metal contacts on ZnO changed with time or after annealing [9, 10]. The effects of the interface oxidization on electrical properties have been discussed [11]. Both pretreatments and metal/ZnO interfaces are essential for realizing a high-quality Schottky contact. Furthermore, the Schottky barrier height (SBH) is related to the work function of materials. High-density interface defects reduce the metal work function [12], suggesting that an understanding of the metal/ZnO interfaces is the key to controlling the properties of Schottky contacts.

ZnO has a wurtzite crystal structure with two distinct {0001} planes, as shown in Fig. 2.1. A lack of inversion symmetry and ionic bonds make this material polar. The Zn-terminated plane (0001) and the O-terminated plane (000–1) are, respectively, denoted as the Zn-polar face and the O-polar face. These two faces have different structures, compositions, and chemical and physical properties [13, 14]. The interface structure and thermal stabilities of metal/ZnO substrates should therefore be different.

© National Institute for Materials Science, Japan 2020
T. Nagata, *Nanoscale Redox Reaction at Metal/Oxide Interface*, NIMS Monographs,
https://doi.org/10.1007/978-4-431-54850-8_2

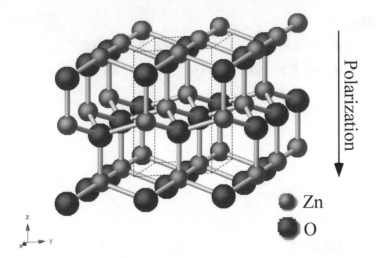

Fig. 2.1 Schematic illustration of the crystal structure of ZnO

Here, we report on the SBH behavior of Pt–Ru alloy contacts on n-ZnO substrates at both the Zn-polar and O-polar faces. Pt and Ru are, respectively, used as high and low work function materials. Furthermore, the Gibbs free energy of Pt oxide is larger than that of Ru oxide, indicating that the Ru oxide layer is more easily formed. Using the combination of polarity, work function, and oxidization energy differences, we investigated the interface structure effects on the SBH behavior by the combinatorial synthesis technique.

2.2 Interface Formation and Characterization

2.2.1 Schottky Barrier Height

Figure 2.2 shows the energy-level diagram for a metal and *n*-type oxide semiconductor before and after contact formation. At the metal/semiconductor interface, in the absence of defect states, Schottky barrier height (Φ_B) is given by

$$\Phi_B = \Phi_M - \Phi_S, \tag{2.1}$$

where Φ_M is the metal work function and Φ_S is the semiconductor work function. Φ_B is also given by

$$\Phi_S = \chi + (E_C - E_F), \tag{2.2}$$

(a)   **(b)**

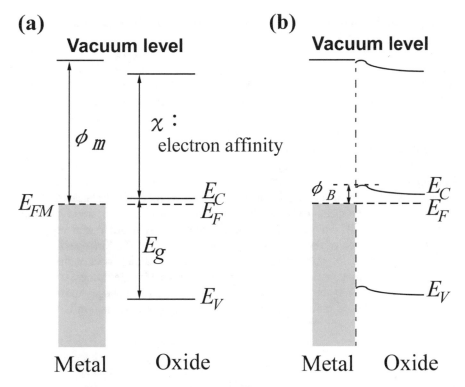

Fig. 2.2   Schematic energy-level diagrams for **a** an ideal metal/oxide interface prior to contacting, and **b** metal contact on an oxide film. E_{FM}: Fermi-level of the metal, E_F: Fermi-level of the oxide, E_C: conduction band, E_V: valence band, Eg: band-gap, Φ_m: work function of the metal, X: electron affinity of the oxide

with the electron affinity χ, the Fermi-level E_F, and the conduction band edge E_C. For an n-type semiconductor, the Fermi-level is close to the conduction band, and for $\Phi_S \approx \chi$, we approximate

$$\Phi_B = \Phi_M - \chi. \tag{2.3}$$

However, an actual metal/oxide interface shows a smaller Φ_B than that calculated using Eq. (2.1). The metal/oxide interface is affected by various phenomena, such as oxidization and defect formation.

2.2.2   Gibbs Free Energy: Ellingham Diagram

To use oxides and metal/oxide heterostructures, the Gibbs free energy (ΔG) is useful. Gibbs free energy is a thermodynamic potential that measures the thermodynamic driving force that makes a reaction (oxidization) to occur. ΔG is given by

$$\Delta G = \Delta H - T\Delta S \tag{2.4}$$

where ΔH is the enthalpy, T is the absolute temperature, and ΔS is the entropy. ΔH is the actual energy. If it is negative, the reaction is exoergic. In contrast, if it is positive, the reaction is endoergic. ΔS is a measure of the disorder of a system. In the case of oxide formation from a metal, a negative value for ΔG means that oxidization can proceed spontaneously without external inputs.

The standard Gibbs free energy of formation of a compound (oxide) is the change in Gibbs free energy that accompanies the formation of 1 mol of a substance in its standard state from its constituent elements in their standard states (the most stable form of the element at 1 bar of pressure and the specified temperature, usually 298.15 K or 25 °C). An Ellingham diagram plots the standard free energy of a reaction as a function of temperature, as shown in Fig. 2.3 [15]. Since ΔH and ΔS are essentially constant with temperature unless a phase change occurs, the plot of free energy versus temperature can be drawn as a series of straight lines, where ΔS is the slope and ΔH is the y-intercept. The slope of the line changes when any of the materials involved melt or vaporize. The free energy of formation is negative

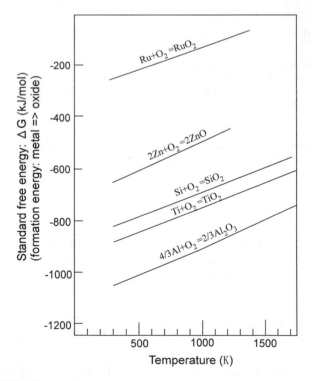

Fig. 2.3 Ellingham diagram

for most metal oxides, and so the diagram is drawn with $\Delta G = 0$ at the top, and the values of ΔG shown are all negative numbers. The Ellingham diagram shown is for metals reacting to form oxides. The oxygen partial pressure is taken as 1 atmosphere, and all the reactions are normalized to consume 1 mol of O_2. By using the diagram, the standard free energy change for any included reaction can be found at any temperature. Along with enabling the calculation of the equilibrium composition of the system, the data on the diagram is useful in other ways, as we shall see.

2.2.3 Phase Diagram

The SBH is related to the work function of the metal. It is generally known that the work function of a metal is affected by the surface potential, which means an effect of the orientation of the metal layer, such as tungsten [16–18]. In the case of tungsten, the values of the work functions of (111), (100), and (110) faces are 4.47, 4.63, and 5.25 eV, respectively. Additionally, the work function of a metal is affected by interface defects. A high density of defects reduces the value of the work function [19]. To control the Schottky barrier height, the crystal structures of the Schottky metal layer should be investigated. For understanding and controlling the crystal structure of alloys, a phase diagram is beneficial. In the case of Pt–W as shown in Fig. 2.4, there is no intermediate stoichiometric alloy, meaning that the Pt–W alloy has a possibility of linear controllability of the work function. Actually, the combinatorial synthesis of the Pt–W alloy revealed the continuously changing behavior of the work function, as shown in Fig. 2.5. The Pt–W alloy has 111-oriented structures. The work function decreases continuously with increasing W content.

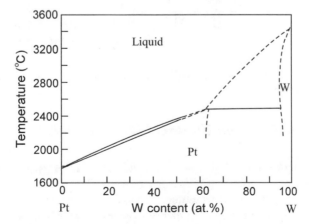

Fig. 2.4 Binary phase diagram of Pt–W alloy

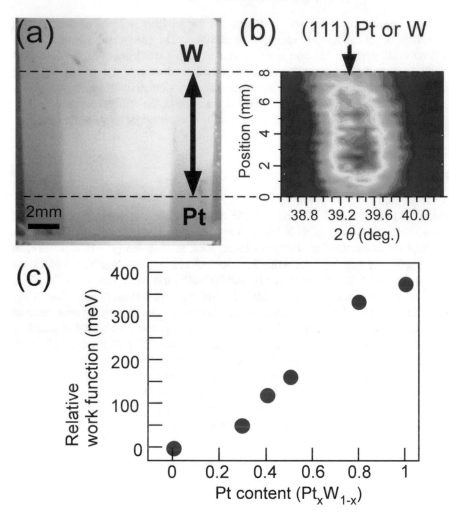

Fig. 2.5 **a** Sample picture, **b** XRD mapping, and **c** relative work functions measured by KFM of Pt–W alloy composition spread sample

2.2.4 Characterization: Electrical Measurements

A typical way to investigate Φ_B is by electrical measurements such as current–voltage (I–V) and capacitance–voltage (C–V) measurements. From I–V measurements, both Φ_B and the ideality factor (n) were determined using Eqs. (2.5) and (2.6), and are consistent with thermionic emission theory in excess of several kT/q [20].

$$I = I_0 \exp\left[\frac{q(V - IR_s)}{nkT} - 1\right], \qquad (2.5)$$

$$I_0 = AA^{**}T^2 \exp\left[\frac{-q\Phi_B}{kT}\right], \tag{2.6}$$

where I_0 is the saturation current, R_s is the series resistance, k is Boltzmann's constant, T is the absolute temperature, A is the contact area, q is the electronic charge, and A^{**} is the effective Richardson constant.

From C–V measurements, the majority carrier distribution $n(x)$ is given by [21, 22]

$$n(x) = -\frac{C^3}{qk\varepsilon_0}\left(\frac{dC}{dV}\right)^{-1}, \tag{2.7}$$

where $k\varepsilon_0$ is the semiconductor permittivity, q is the electron charge, and $x = k\varepsilon_0 A/C$ is the test junction space charge layer width at the applied biasing voltage V. Using the majority carrier distribution, Φ_B can be estimated.

2.2.5 Characterization: X-Ray Photoelectron Spectroscopy

X-ray photoelectron spectroscopy (XPS) is an electron spectroscopy method. X-rays eject electrons from inner-shell orbitals of the elements that exist within a material; then, an energy analyzer detects the photoelectrons. The kinetic energy (E_K) of the photoelectrons is determined by the energy of the X-ray ($h\nu$) and the electron binding energy (E_B) as

$$E_K = h\nu - E_B. \tag{2.8}$$

The experimentally measured binding energy is given by

$$E_B = h\nu - E_K - \varphi_{spec}, \tag{2.9}$$

where φ_{spec} is the work function of the spectrometer. The electron binding energies are dependent on the chemical environment of the atom, making XPS useful for identifying the chemical bonding state (oxidation state) and electronic state of materials.

The X-ray penetration depth is at least several micrometers. However, the electrons leave with very low energy, so they can only escape from the top few nanometers. Tanuma, Powell, and Penn developed the Tanuma–Powell–Penn (TPP-2 M) equation to calculate the inelastic mean free paths (IMFPs: λ) using Green's function [23]. The TPP-2 M equation indicated that IMFP can become longer by using high-energy X-rays. In our study, a combination of conventional X-ray photoelectron spectroscopy using Al $K\alpha$ radiation (Al-XPS: $h\nu = 1486.6$ eV) and hard X-ray photoelectron spectroscopy (HX-PES: $h\nu = 5.95$ keV) was used to investigate the

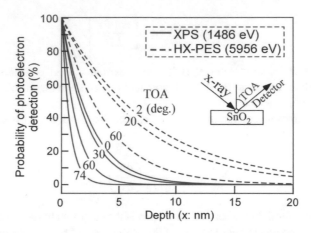

Fig. 2.6 Probability of photoelectron detection of SnO_2 based on IMFP calculation at various take-off angles (TOAs). The solid and dashed lines show conventional XPS and HX-PES, respectively. The inset shows the measurement setup

interface structure. For example, the IMFPs for the Sn $3d_{5/2}$ core-level spectra of Al-XPS and HX-PES for SnO_2 calculated using the TPP-2 M equation are 1.8 and 6.9 nm, respectively, indicating that HX-PES probes three times deeper than Al-XPS. Furthermore, to determine the details of band bending behavior, angle-resolved XPS (AR-XPS) measurements were carried out. While crossing the material, the photo-electrons are subject to the laws of absorption. Thus, the probability (I_0) that the emitted photoelectron reaches the surface can be estimated against a given depth (x) of photoemission. Figure 2.6 shows an example of the probability ($I_0 = \exp(-x/\lambda)$) for the Sn $3d_{5/2}$ core-level spectra at various measurement angles as a function of depth. This plot indicates that, by changing the measurement angle, the probing depth can be controlled and it is approximately $3 \times \lambda$ [24, 25].

In our experiments, HX-PES was performed at the SPring-8 BL15XU undulator beamline with a 200 mm mean radius spectrometer (VG Scienta R4000) [26]. The total energy resolutions for the Al-XPS and HX-PES measurements were estimated to be ~600 and ~230 meV, respectively. To determine the absolute binding energy, the photoelectron spectroscopy data was calibrated against the Fermi-level position of Au. The sample was in contact with the system ground, whose energy was equal to the Fermi-level position of Au, via a conductive copper tape. To analyze the HX-PES results, peaks were fitted using the Voigt function with the Doniach-Šunjić function after the background had been removed by employing the Shirley function [27, 28].

At the site of metal electrode formation, XPS can describe the band offset and bending of a metal/oxide interface. Since the core levels have a fixed binding energy difference from the conduction and valence band edges, they can be used to trace shifts of the band edges with respect to the Fermi level, as shown in Fig. 2.7. Fur-thermore, HX-PES probes at 20 nm below the surface, indicating the bulk property of ZnO. Figure 2.8 shows the Zn $2p_{3/2}$ spectra of the Zn-polar ZnO substrate and 10 nm Pt/ZnO interface obtained by HX-PES. The Pt electrode formation moved

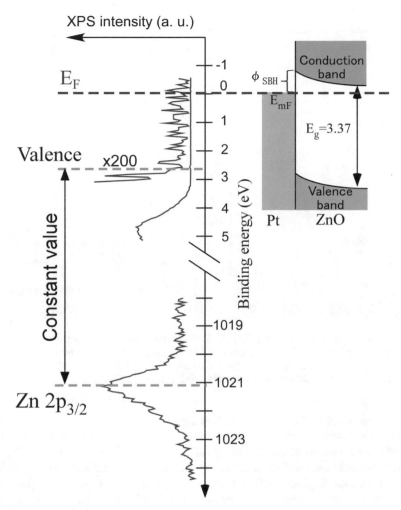

Fig. 2.7 Correlation of valence band and core spectra in photoelectron spectroscopy and energy band diagram of Pt/ZnO interface

Zn $2p_{3/2}$ to a lower binding energy, meaning that the Fermi-level moved to the middle of the band-gap of ZnO at the interface. Furthermore, the take-off angle (TOA) dependence of the Zn $2p_{3/2}$ position revealed the upward band bending of the ZnO at the interface, shown in Fig. 2.7, suggesting the formation of a depletion region at the ZnO surface. These results are consistent with Schottky contact formation at the metal/semiconductor interface.

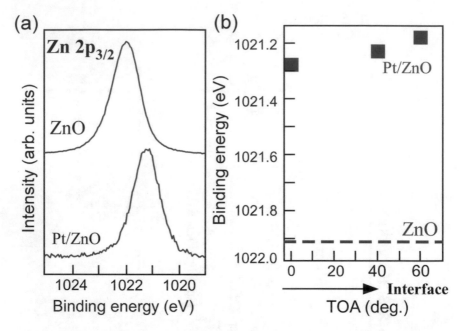

Fig. 2.8 **a** Zn $2p_{3/2}$ core spectra of a Zn-polar ZnO substrate and a Pt/ZnO interface at a TOA of 88°. **b** TOA dependence of the binding energy of the Zn $2p_{3/2}$ core spectra. The squares show the Pt/ZnO interface. The Zn $2p_{3/2}$ core spectrum of a Zn-polar ZnO substrate at a TOA of 88°, which probes a similar depth to the others in terms of the IMFP, is also indicated as a dashed line

For measurements of metal work function, photoelectron spectroscopy is also useful. Under even atmospheric pressure conditions, small numbers of low-energy photoelectrons can be detected. This photoelectron yield spectroscopy (PYS) can estimate not only the yield spectrum but also DOS as UPS and XPS, in air ambient. Figure 2.9 shows a photoelectron yield spectrum of the Ru film on a glass substrate. The ionization energy is defined as the energy difference between the vacuum level and the extrapolated edge of the Fermi-level of the metal. Therefore, the work function can be estimated by linearly extrapolating the edge of the square-root photoelectron yield to the baseline. The work function of the Ru film is 4.85 ± 0.1 eV, which is comparable to previously reported values (4.6–4.7 eV) [16–18].

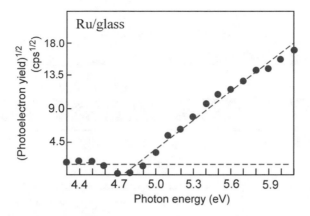

Fig. 2.9 Photoelectron yield spectrum of a Ru film on a glass substrate

2.3 Combinatorial Synthesis of Binary Alloy Metal Contacts on Polar Face of ZnO

Figure 2.10a shows a schematic of the sample structure fabricated by combinatorial synthesis technique. The details of combinatorial synthesis technique are shown in Chap. 6. An ohmic contact layer was made by the Al lift-off method. Circular top Schottky contacts of 130 μm diameter were deposited by the composition spread method. Pt and Ru were deposited as Schottky metals. Figure 2.10b shows the PSY measurement of a Pt–Ru alloy on a glass substrate. The work functions of the Pt–Ru composition spread sample change continuously.

2.3.1 Crystal Structural Analysis

The 2D XRD images of the Pt and Ru films showed spotted peaks at approximately 40° (Fig. 2.11a). The peaks were assigned to a cubic structure with (111) reflection (Pt) and a hexagonal structure with (0002) reflection (Ru).

The peaks of the electrodes are observed at approximately 40°. These peaks shift to higher angles with increasing Ru content and have two possible crystal structures. One is the Pt phase with a cubic structure, and the other structure is the Ru phase with a hexagonal structure. The electrodes with a Pt-type structure show only the (111) reflection, indicating a 111-oriented film. Those with a Ru-type structure show only the (0002) reflection, indicating a 0001-oriented film. The Pt–Ru alloy phase diagram has two intermediate phase lines of crystal structures changing at Pt content of approximately 20 and 39 at.%, as shown in Fig. 2.11b [29]. At a Pt content below 20 at.%, the crystal structure of the Pt–Ru alloy is hexagonal; at ≥39 at.%, it is cubic; between 20 and 39 at.%, it is a mixed structure phase. The rocking curves of the Pt

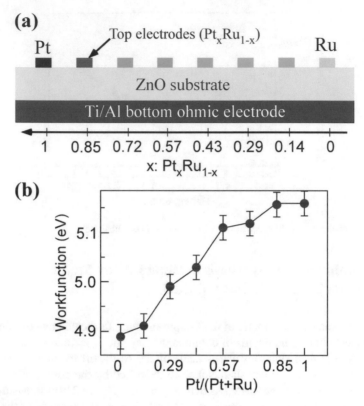

Fig. 2.10 a Schematic illustration of the Pt–Ru composition spread sample. **b** Work functions measured by PYS of the Pt–Ru composition spread film on a glass substrate

phase with cubic structure (111) reflections or the Ru phase with hexagonal structure (0002) reflections are plotted as a function of Pt content, as shown in Fig. 2.11c. FWHM increased with decreasing Pt content, and showed a peak at a Pt content of 29 at.%. This change is suitable for the Pt–Ru alloy phase diagram, because the Pt content of 29 at.% results in a mixed structure phase. Mixed structure phases may become a low crystalline phase.

From the results of the ω-2θ XRD and ψ scans, it cannot be determined whether the film is of the Pt phase or the Ru phase. X-ray pole figure analysis was also performed. The X-ray pole figure was measured at 3° intervals of scan step in ϕ angle. The ranges of the 2θ and ψ angles are detected simultaneously from 30 to 60° and from 25 to 75°, respectively. Figure 2.12a–c shows the results of Pt content of 0, 29, and 100 at.%, respectively. The value of ψ was 42.7° for the $(10\bar{1}2)$ plane of ZnO. In Fig. 2.12a, peaks with a six-fold symmetry at 60° in the ϕ scan can be observed at $\psi = 61°$ for the $(10\bar{1}1)$ plane of Ru. This implies that Ru with a hexagonal structure grew on the ZnO substrate with its c-axis normal to the substrate surface epitaxially. Figure 2.12c shows peaks with a six-fold symmetry at $\psi = 54.7°$

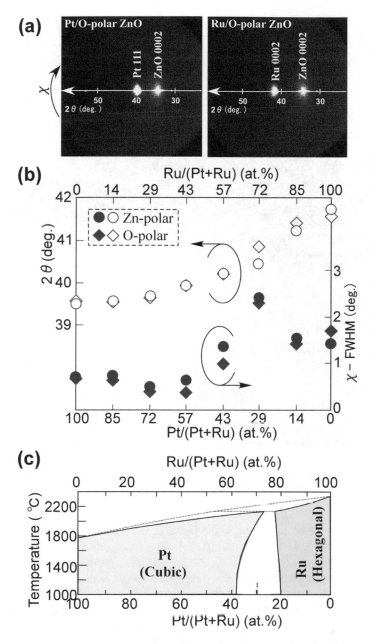

Fig. 2.11 **a** 2D-XRD images of the Pt and Ru films on the O-polar face. **b** Dependence of peak positions and χ-FWHMs on Pt content for Pt (111) reflection with a cubic structure or Ru (0002) reflection with a hexagonal structure on Zn-polar (solid squares) and O-polar (open squares) faces. **c** Pt–Ru alloy phase diagram

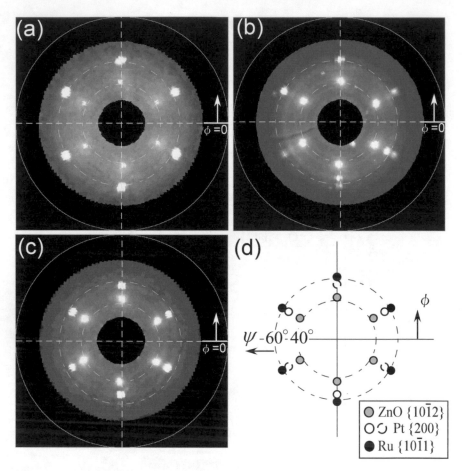

Fig. 2.12 X-ray pole figures of **a** Ru, **b** Pt content of 29 at.%, **c** Pt electrodes on ZnO, and **d** theoretical pole figures of Ru and Pt on ZnO

for the (200) plane of Pt, although the peaks of the 111-oriented epitaxial Pt film showed a normal three-fold symmetry. This result suggests that the Pt phase with a rotation of 60° was also grown on ZnO epitaxially. At the point of 29 at.% on the mixed structure phase, the X-ray pole figure shows a mixed pattern of Ru and Pt phases. These results reveal that the crystal structures of the Pt–Ru alloy film on ZnO change from the Pt phase to the Ru phase at a Pt content of approximately 29 at.%, as in the case of the Pt–Ru alloy phase diagram, and that the Pt–Ru alloy on a ZnO was grown on the ZnO substrate epitaxially.

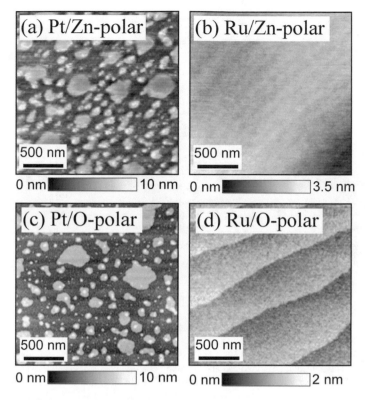

Fig. 2.13 AFM images of **a** Pt and **b** Ru films on Zn-polar face, and **c** Pt and **d** Ru films on O-polar face

2.3.2 Surface Morphology

The Pt–Ru alloy film surfaces were observed by AFM, as shown in Fig. 2.13. Both polar faces showed similar surface changes. The Pt films had small-grained structures with a root mean square (RMS) value of 1.72 nm at the Zn-polar face. In contrast, the low-Pt films showed flat surfaces with an RMS value of 0.18 nm for Ru at the Zn-polar face. The maximum RMS difference between Zn-polar and O-polar faces was 0.07 nm. Thus, the surface roughness difference should have a negligible effect on the surface potential.

2.3.3 Electrical Properties

The I–V measurements revealed that the Schottky properties become dominant with increasing Pt content of the films. Below 13 at.%, in the Ru structure phase, ohmic behaviors become dominant. To investigate SBH, we modeled the I–V characteristics

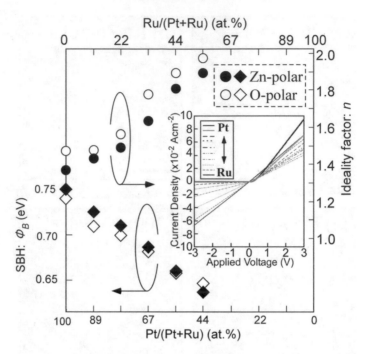

Fig. 2.14 Pt content dependences of SBH (Φ_B: squares) and ideality factor (n: circles) of the Pt–Ru alloy films on Zn-polar (solid marks) and O-polar (open marks) faces calculated from the I–V characteristic. The inset shows the I–V characteristic of Pt–Ru alloy contact on the O-polar face

of all Pt–Ru alloy contacts using Eqs. (2.4) and (2.5). In this study, the theoretical value of $A^{**} = 32$ A cm^{-2} K^{-2} was used [23]. Figure 2.14 shows the dependence of SBH and the ideality factor on Pt content. At both polar faces, SBH decreased with decreasing Pt content, suggesting that the metal work function changed continuously. SBH changed by 0.11 eV at the Zn-polar face and by 0.09 eV at the O-polar face. SBH was greater at the Zn-polar face than at the O-polar face except at a Pt content of 44 at.%. The ideality factors on the O-polar face are larger than that on the Zn-polar face. Furthermore, with increasing Ru content, the ideality factor increases. High ideality factor means an increase of the recombination current instead of the ideal diffusion current, suggesting the defect formation at the interface.

2.3.4 Chemical Bonding States (HX-PES Measurements)

To investigate the electronic states of the Pt–Ru/ZnO interface, we performed HX-PES measurements at different compositions of 10 nm thick Pt–Ru alloy films on the Zn-polar and O-polar faces. Figure 2.15a shows the Pt 4f and Ru 3d core-level HX-PES spectra for Pt and Ru at the O-polar face at a TOA of 88° from the surface.

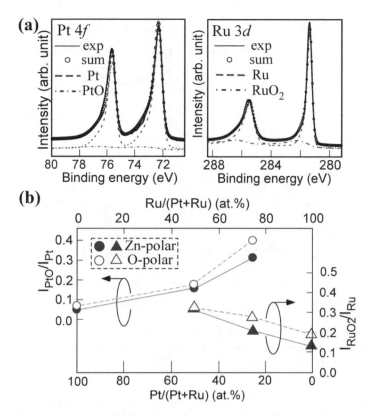

Fig. 2.15 **a** Typical Pt 4f and Ru 3d core-level HX-PES spectra of Pt and Ru on the O-polar face at a TOA of 88°. Solid lines, measured spectra; open circles, sum-fitted curves. Dashed lines are fitted curves for each bond: Pt, PtO, Ru, and RuO_2. **b** Dependence of the HX-PES spectrum intensity ratio of oxide to metal on Pt and Ru content (circles, PtO/Pt; triangles, RuO_2/Ru) at a TOA of 88°. Solid symbols, Pt–Ru alloy films on Zn-polar faces; open symbols, on O-polar faces

In the Pt film, the peaks at the binding energies of 72.3 and 75.6 eV correspond to metallic Pt, and small peaks at 73.5 and 76.8 eV are attributed to PtO [30, 31]. In the Ru film, peaks at 281.3 and 285.5 eV correspond to metallic Ru, and small peaks at 282.5 and 286.7 eV are attributed to RuO_2 [32, 33]. Figure 2.15b shows the relative intensities of the metals and the oxides as a function of the Pt or Ru content. The intensity ratio of the films was higher on the O-polar face than on the Zn-polar face. These results suggest that the Pt–Ru alloy films on the O-polar face were more oxidized. With decreasing Pt content, the PtO/Pt ratio increases, and the RuO_2/Ru ratio decreases slightly, indicating that the alloy film, which is in a mixed phase and poorly crystalline, has the most oxidized layer in the Pt–Ru alloy films.

At a TOA of 20°, a clear difference in the Zn $2p_{3/2}$ spectrum was confirmed, as shown in Fig. 2.16. Intensities of the Zn $2p_{3/2}$ spectra were normalized to the corresponding Pt or Ru core-level peak intensities. The Zn $2p_{3/2}$ intensity of the Zn-polar face is much greater than that of the O-polar face, suggesting that the Zn

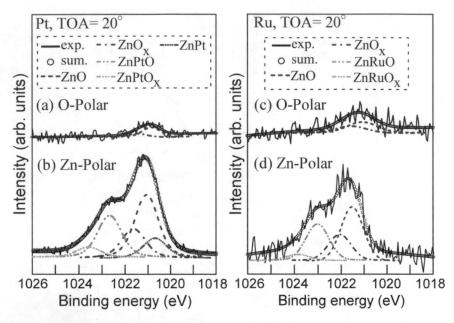

Fig. 2.16 (Color online) Zn 2p$_{3/2}$ spectra of Pt and Ru films on O-polar and Zn-polar faces at a TOA of 20°. Solid lines, measured spectra; open circles, sum-fitted curves. Dashed lines are fitted curves for each bond: ZnO, ZnO$_x$, ZnPt, ZnPtO, ZnPtO$_x$, ZnRuO, and ZnRuO$_x$

in the Zn-polar face is present at a shallower location than in the O-polar face. To confirm this consideration, javascript:goWordLink(%22maintenance%22) the IMFP of photoelectrons was calculated using the TPP-2 M equation. The Pt/ZnO interface structure was modeled using two different structures. One was PtO for the Pt/O-polar ZnO interface, and the other was PtZn for the Pt/Zn-polar interface. The IMFPs of PtZn and PtO are 5.26 and 6.11 nm, respectively. The IMFP of PtO is longer than that of PtZn. If the top of Zn layers of both polar faces were present at a similar depth within the difference of one monolayer length, the intensity of the Zn 2p for PtO would appear greater than that for PtZn. This means that the Zn in the Zn-polar face is present at a shallower location than in the O-polar face, indicating that Zn diffused into the Pt layer. Moreover, unlike the O-polar face, three additional peaks were identified in the Zn-polar face, as shown in Fig. 2.16b: ZnPt, ZnPtO, and ZnPtOx. These species were assigned considering the difference in electronegativity of each atom and in reference to previous studies [34, 35]. Some groups reported that the Zn 2p peak of the ZnO bond of zinc ferrite and zinc titanate, in which the coordination number of Zn is six, shifted to a higher binding energy than the Zn 2p peak of four coordinate Zn atoms in oxides. In our case, ZnPt and ZnPt alloys below a Pt content of 48% have a cubic structure in which the coordination number of Zn is six [36]. Furthermore, some of the ZnPt area is likely to be oxidized. The Zn in oxidized ZnPt, such as Zn$_2$PtO$_4$, also has a coordination number of six. Thus, the peaks at higher binding energy were assigned as ZnPtO and ZnPtOx bonds. For the Ru film, the component at the lowest binding energy corresponding to a metal–Zn bond was not identified. The Ru film was more oxidized than the Pt film.

2.3.5   Surface Termination Effect

The metal work functions of Pt and Ru are 5.32–5.50 and 4.6–4.7 eV, respectively, and the electron affinity of ZnO is 4.1–4.4 eV [37–39]. Thus, SBH between the Pt–Ru alloy and ZnO should be in the range of 1.1–0.3 eV. Furthermore, the Zn-polar face appears to show a slightly higher SBH than the calculated value, owing to polarization; this bends the band edge upward. The O-polar face should show the opposite property.

In the I–V measurement, the SBH increased with increasing Pt content, and the Zn-polar face showed a slightly higher SBH than the O-polar face. However, the SBHs were smaller than the calculated values, and the films with the Ru phase showed ohmic properties. According to the crystal structural analysis and HX-PES results, these phenomena can be explained by the complex interaction of three factors: (A) oxidization, (B) Zn- or O-termination, and (C) crystal structure of the films on the SBH.

(A) Oxidization: In the Ru phase, the alloys on both polar faces of ZnO showed ohmic properties, and the O-polar face showed poor electrical properties. These phenomena were caused by the oxidization of the metal layer. The HX-PES results indicate that the Pt–Ru film was more oxidized on the O-polar face than on the Zn-polar face. In addition, the films were in a mixed phase with poor crystallinity, suggesting that they had the thickest oxidized layer of the Pt–Ru alloy films. Shiraishi et al. reported that oxygen at the interface lowers the metal's Fermi-level, and defects in the interface induced by the interface oxidization decrease the metal's work function [40]. This theoretical result agrees closely with our result.

(B) Zn- or O-termination: The HX-PES results revealed the relationship between the interface structures and the chemical states of the films on the Zn- and O-polar faces. At the Zn-polar face, Zn diffused into the metal layer, which implies a difference in oxidization at the interface. The Zn diffusion was expected to prevent the oxidization of metal, since Zn is used as a deoxidant in alloys. The Zn-polar face appears to inhibit the oxidization of the contact more than the O-polar face. However, the work function of Pt (5.32–5.50 eV) is likely to decrease owing to the metal work function of Zn (3.74 eV). Therefore, SBH decreases. At the O polar face, in contrast, metal oxide species form thickly around the interface. Thus, the SBH of the O-polar face is smaller than that of the Zn-polar face.

(C) Crystal structure: The changes in SBH of Pt–Ru alloy electrodes on the Zn-polar and O-polar faces were 0.11 and 0.09 eV, respectively. The difference in SBH between the highest and lowest values is smaller than the value calculated from the work function. If the change in SBH is continuous, the difference in SBH from the Pt film to the Pt–Ru film with a Pt content of 44 at.% should be 0.44 eV. However, SBH decreased slightly (Fig. 2.15) and changed the ohmic contact at the Ru phase. Some groups reported that the metal work function changing behavior of alloys shows some dependence on the phase of the alloy [41, 42]. The XRD and ellipsometric measurements indicate that the crystal structure affects the electrical properties: the Pt–Ru alloy films with Schottky behaviors showed the cubic structure (Pt phase) and the same optical properties at a Pt content of >39 at.%.

2.4 Summary

Pt–Ru alloy composition spread films were deposited on ZnO single-crystal substrates as Schottky contacts by combinatorial ion-beam deposition. The crystal structures of the Pt–Ru alloy changed from cubic to hexagonal phase with increasing Pt content, as seen in the bulk Pt–Ru alloy phase diagram. The SBH of the films in the Pt phase decreased with decreasing Pt content irrespective of the polar face of the ZnO substrate, although the Pt–Ru alloy with a Pt content of <13 at.% did not show Schottky behavior. Furthermore, the O-polar face showed poorer Schottky properties than the Zn-polar face. The HX-PES measurements revealed Zn diffusion and interface oxidization. At the Zn-polar face, the diffused Zn inhibited metal oxidization and reduced the metal work function. The O-polar face and the mixed phase with a Pt content of 20–39 at.% had a more oxidized layer than the Zn-polar face or other phases. Owing to these oxidization layers, the SBH was smaller than the theoretical value, and Ru showed ohmic properties.

In conclusion, Pt–Ru alloy contacts on n-ZnO were affected by oxidization, Zn- or O-termination, and crystal structure. To control the SBH of metals on ZnO, the key factors are alloying and the control of oxidization of the metal/ZnO interface.

References

1. Bagnall DM, Chen YF, Zhu Z, Yao T, Koyama S, Shen MY, Goto T (1997) Optically pumped lasing of ZnO at room temperature. Appl Phys Lett 70:2230. https://doi.org/10.1063/1.118824
2. Tsukazaki A, Ohtomo A, Onuma T, Ohtani M, Makino T, Sumiya M, Ohtani K, Chichibu SF, Fuke S, Segawa Y, Ohno H, Koinuma K, Kawasaki M (2005) Repeated temperature modulation epitaxy for p-type doping and light-emitting diode based on ZnO. Nat Mater 4:42–46. https://doi.org/10.1038/nmat1284
3. Pearton SJ, Norton DP, Ip K, Heo YW, Steiner T (2001) Recent advances in processing of ZnO. J Vac Sci Technol B 22:932. https://doi.org/10.1116/1.1714985
4. Look DC (2001) Recent advances in ZnO materials and devices. Mater Sci Eng B 80:383–387. https://doi.org/10.1016/S0921-5107(00)00604-8

5. Nagata T, Ahmet P, Chikyow T (2007) Crystal structures of Pt–Ru alloy Schottky contacts on ZnO by combinatorial ion beam deposition. Jpn J Appl Phys Part 1(46):2907–2909. https://doi.org/10.1143/JJAP.46.2907

6. Oba F, Nishitani S, Isotani S, Adachi H, Tanaka I (2001) Energetics of native defects in ZnO. J Appl Phys 90:824. https://doi.org/10.1063/1.1380994

7. Allen MW, Durbin SM, Metson JB (2007) Silver oxide Schottky contacts on n-type ZnO. Appl Phys Lett 91:053512. https://doi.org/10.1063/1.2768028

8. Endo H, Sugibuchi M, Takahashi K, Goto S, Sugimura S, Hane K, Kashiwaba Y (2007) Schottky ultraviolet photodiode using a ZnO hydrothermally grown single crystal substrate. Appl Phys Lett 90:121906. https://doi.org/10.1063/1.2715100

9. Kim HK, Han SH, Seong TY, Choi WK (2000) Low-resistance Ti/Au ohmic contacts to Al-doped ZnO layers. Appl Phys Lett 77:1647. https://doi.org/10.1063/1.1308527

10. Allen MW, Alkaisi MM, Durbin SM (2006) Metal Schottky diodes on Zn-polar and O-polar bulk ZnO. Appl Phys Lett 89:103520. https://doi.org/10.1063/1.2346137

11. Allen MW, Durbin SM (2008) Influence of oxygen vacancies on Schottky contacts to ZnO. Appl Phys Lett 92:122110. https://doi.org/10.1063/1.2894568

12. Besocke K, Krahl-Urban B, Wagner H (1977) Dipole moments associated with edge atoms; A comparative study on stepped Pt, Au and W surfaces. Surf Sci 68:39–46. https://doi.org/10.1016/0039-6028(77)90187-X

13. Matsui H, Saeki H, Kawai T, Sasaki A, Yoshimoto M, Tsubaki M, Tabata H (2004) Characteristics of polarity-controlled ZnO films fabricated using the homoepitaxy technique. J Vac Sci Technol, B 22:2454. https://doi.org/10.1116/1.1792237

14. Ohashi N, Adachi Y, Ohsawa T, Matsumoto K, Sakaguchi I, Haneda H, Ueda S, Yoshikawa H, Kobayashi K (2009) Polarity-dependent photoemission spectra of wurtzite-type zinc oxide. Appl Phys Lett 94:122102. https://doi.org/10.1063/1.3103271

15. Ellingham HJT (1944) Reducibility of oxides and sulphides in metallurgical processes. J Soc Chem Ind (London) 63:125. https://doi.org/10.1002/jctb.5000630501

16. Michaelson HB (1977) The work function of the elements and its periodicity. J Appl Phys 48:4729. https://doi.org/10.1063/1.323539

17. Krahl-Urban B, Niekisch EA, Wagner H (1977) Work function of stepped tungsten single crystal surfaces. Surf Sci 64:52–68. https://doi.org/10.1016/0039-6028(77)90257-6

18. Venbles JA (2000) Introduction to surface and thin film process. Cambridge University Press, Cambridge, p 194

19. Besocke K, Krahl-Urban B, Wagner H (1997) Dipole moments associated with edge atoms; A comparative study on stepped Pt, Au and W surfaces. Surf Sci 68:39–46. https://doi.org/10.1016/0039-6028(77)90187-X

20. Sze SM (1981) Physics of semiconductors devices. Wiley, New York, p 849

21. Kennedy DP, O'brien RR (1969) On the measurement of impurity atom distributions by the differential capacitance technique. IBM J. Res. Dev 13:212. https://doi.org/10.1147/rd.132.0212

22. Rickert KA, Ellis AB, Himpsel FJ, Lu H, Schaff W, Redwing JM, Dwikusuma F, Kuech TF (2003) X-ray photoemission spectroscopic investigation of surface treatments, metal deposition, and electron accumulation on InN. Appl Phys Lett 82:3254. https://doi.org/10.1063/1.1573351

23. Tanuma S, Powell CJ, Penn DR (1988) Calculations of electron inelastic mean free paths for 31 materials. Surf Interface Anal 11.577–589. https://doi.org/10.1002/sia.740111107

24. Powell CJ, Jablonski A, Tilinin IS, Tanuma S, Penne DR (1999) Surface sensitivity of auger-electron spectroscopy and X-ray photoelectron spectroscopy. J Electron Spectrosc Relat Phenom 98–99:1–15. https://doi.org/10.1016/S0368-2048(98)00271-0

25. Tanuma S (2006) Electron scattering effect on surface electron spectroscopies. J Surf Sci Soc Japan 27:657. https://doi.org/10.1380/jsssj.27.657

26. Ueda S, Katsuya Y, Tanaka M, Yoshikawa H, Yamashita Y, Matsushita Y, Kobayashi K (2010) Present status of the NIMS contract beamline BL15XU at SPring-8. AIP Conf Proc 1234:403. https://doi.org/10.1063/1.3463225

27. Doniach S, Šunjić M (1970) Many-electron singularity in X-ray photoemission and X-ray line spectra from metals. J Phys C 3:285. https://doi.org/10.1088/0022-3719/3/2/010
28. Shirley DA (1972) High-resolution X-ray photoemission spectrum of the valence bands of gold. Phys Rev B 5:4709. https://doi.org/10.1103/PhysRevB.5.4709
29. Hutchinson JM Jr (1972) Solubility relationships in the ruthenium-platinum system. Platinum Met Rev 16:88–90
30. Bancroft GM, Adams I, Coatsworth LL, Bennewitz CD, Brown JD, Westwood W (1975) ESCA study of sputtered platinum films. Anal Chem 47:586–588. https://doi.org/10.1021/ac60353a050
31. Barr TL (1978) An ESCA study of the termination of the passivation of elemental metals. J Phys Chem 82:1801–1810. https://doi.org/10.1021/j100505a006
32. Kim KS, Winograd N (1974) X-ray photoelectron spectroscopic studies of ruthenium-oxygen surfaces. J Catal 35:66–72. https://doi.org/10.1016/0021-9517(74)90184-5
33. Lewerenz HJ, Stucki S, Kötz R (1983) Oxygen evolution and corrosion: XPS investigation on Ru and RuO₂ electrodes. Surf Sci 126:463–468. https://doi.org/10.1016/0039-6028(83)90744-6
34. Bera S, Prince AAM, Velmurugan S, Raghavan PS, Gopalan R, Panneerselvan G. Narashimhan SV (2001) Formation of zinc ferrite by solid-state reaction and its characterization by XRD and XPS. J Mater Sci 36:5379–5384. https://doi.org/10.1023/A:1012488422484
35. Druska P, Steinike U, Šepelák V (1999) Surface structure of mechanically activated and of mechanosynthesized zinc ferrite. J Solid State Chem 146:13–21. https://doi.org/10.1006/jssc.1998.8284
36. Massalski TB, Okamoto H, Subramanian PR, Kacprzac L (eds) (1990) Binary alloy phase diagrams, vol 3, 2nd edn. ASM International, OH, p 3153
37. Chen ECM, Wentworth WE, Ayala JA (1977) The relationship between the Mulliken electronegativities of the elements and the work functions of metals and nonmetals. J Chem Phys 67:2642. https://doi.org/10.1063/1.435176
38. Michaelson HB (1950) Work functions of the elements. J Appl Phys 21:536. https://doi.org/10.1063/1.1699702
39. Aranovich JA, Golmayo D, Fahrenbruch AL, Bube RH (1980) Photovoltaic properties of ZnO/CdTe heterojunctions prepared by spray pyrolysis. J Appl Phys 51:4260. https://doi.org/10.1063/1.328243
40. Shiraishi K, Akasaka Y, Miyazaki S, Nakayama T, Nakaoka T, Nakamura G, Torii K, Furutou H, Ohta A, Ahmet P, Ohmori K, Watanabe H, Chikyow T, Green ML, Nara Y, Yamada K (2005) Universal theory of workfunctions at metal/Hf-based high-k dielectrics interfaces—guiding principles for gate metal selection. Technical digest of international electron devices meeting, Washington D.C., USA, pp 39–42. https://doi.org/10.1109/iedm.2005.1609260
41. Bouwman R, Lippits GJM, Sachtler WHM (1972) Photoelectric investigation of the surface composition of equilibrated Ag·Pd alloys in ultrahigh vacuum and in the presence of CO. J Catal 25:350–361. https://doi.org/10.1016/0021-9517(72)90237-0
42. Franken PEC, Ponec V (1974) Photoelectric work functions of Ni·Al alloys: Clean surfaces and adsorption of CO. J Catal 35:417–426. https://doi.org/10.1016/0021-9517(74)90225-5

Chapter 3
Surface Passivation Effect on Schottky Contact Formation of Oxide Semiconductors

3.1 Introduction

The degradations of Schottky barrier heights (SBHs) and ideality factors (IFs) of a metal/ZnO structure are caused by the formation of oxygen vacancies at the metal/ZnO interface owing to the oxidization of the metal layer, as discussed in this chapter. Such oxidation enables Zn to diffuse into the contact layer, which reduces the work function of the contact [1, 2]. Furthermore, oxide thin films are beset by a crucial problem at the surface in the form of a surface accumulation layer. Several groups have reported the existence of such surface accumulation layers [3, 4], which prevent the formation of a corresponding Schottky contact. We have demonstrated the elimination of the surface accumulation layer, which strongly correlates with oxygen vacancies and/or surface defects, on a SnO_2 thin film by oxygen plasma treatment, resulting in an improvement of SBH and IF [5]. However, the oxygen plasma treatment does not prevent the oxidization of metals, and this interaction reduces the work function of the contact. Thus, we have also proposed the use of nitrogen plasma treatment for the Schottky contact formation of oxide semiconductors. Nitride films and oxy-nitride films are known to resist to oxidation [6], which can be expected to prevent the oxidization of metals and Zn diffusion into the contact layer.

3.2 Near-Atmospheric-Pressure Nitrogen Plasma Treatment

For the nitridation, we employed an atmospheric-pressure plasma (AP) source, which was developed by Yuasa and Yara (Sekisui Chemical Co., Ltd.) [7]. A stable discharge of pure nitrogen plasma is maintained by applying an alternating pulsed voltage between two parallel-plate electrodes at atmospheric pressure without the use of a

© National Institute for Materials Science, Japan 2020
T. Nagata, *Nanoscale Redox Reaction at Metal/Oxide Interface*, NIMS Monographs,
https://doi.org/10.1007/978-4-431-54850-8_3

noble gas and a high-vacuum system. The AP source has the capability of large-area nitridation of an oxide surface at room temperature, which is a nonconventional technique and beneficial to the device fabrication [8].

3.2.1 Atmospheric-Pressure Nitrogen Plasma Source

Atmospheric-pressure pure nitrogen plasma, which has a high plasma density and a low plasma temperature, has potential for large-scale ammonia-free processing and low-temperature growth. It is difficult to generate a stable arcless nitrogen plasma at near-atmospheric pressure. Corona discharging and glow discharging using helium are known to occur in nitrogen plasma at atmospheric pressure. These methods are not suitable for the crystal growth of nitrides owing to the damage caused by the high-gas-temperature plasma and defects brought about by the low nitrogen species density. Yuasa and Yara (Sekisui Chemical Co., Ltd.) have developed a plasma source that maintains stable discharge using various gases such as nitrogen and oxygen without the use of a noble gas by applying an alternating pulsed voltage between two parallel-plate electrodes at atmospheric pressure.

Figure 3.1 shows a schematic of the AP system used in this study. All experiments were carried out in a vacuum chamber with a background pressure of 2×10^{-4} Pa. Two parallel-plate electrodes were separated by a uniform gap of 1 mm. Nitrogen plasma was generated by applying alternating pulsed voltages of 7 kV at a frequency of 30 kHz. Pure nitrogen gas (99.9999%) was controlled by a mass flow controller at 400 sccm. The nitrogen gas was led to the chamber. The discharge pressure was 40 kPa. A c-plane sapphire substrate with an atomically smooth surface was used.

Fig. 3.1 Schematic illustration of the near-atmospheric-pressure plasma system. **a** Direct and **b** remote plasma positions

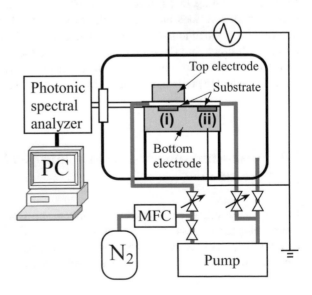

The substrates were placed on the bottom electrode, as indicated in Fig. 3.1 by (i) and (ii). At position (i), the sample was irradiated by the nitrogen plasma directly (denoted "DPS"). Position (ii) was separated from the edge of the electrode by 5 mm, so the sample was irradiated by the plasma remotely (denoted "RPS").

Optical emission spectroscopy, as shown in Fig. 3.2, revealed that the main constituent of the nitrogen plasma was the N_2^+ second positive system (337, 356, 376, and 381 nm) [9]. Herman's infrared system (752, 783, and 806 nm), which is rarely observed in low-pressure plasma and excited atomic nitrogen, was also observed.

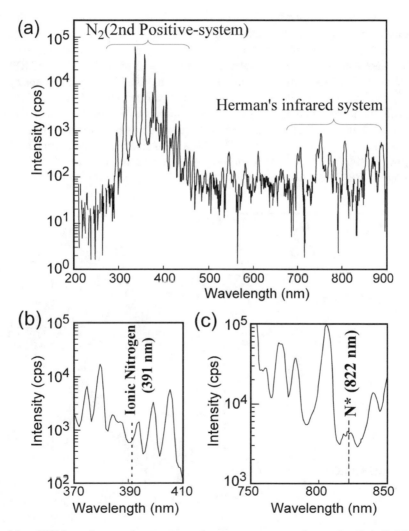

Fig. 3.2 **a** OES from nitrogen plasma generated at room temperature (integration time: 3 s). **b** An enlarged view of OES from nitrogen plasma generated at room temperature from 350 to 450 nm (integration time: 3 s). **c** An enlarged view from 700 to 900 nm (integration time: 20 s)

Additionally, no spectra due to the N_2^+ second positive system, which is an ionic molecular species that causes film damage [10], were observed. This high-density and low-temperature-type nitrogen plasma, including the N_2^+ second positive system, is suitable as a nitridation source.

3.2.2 Nitridation of Oxide Surface

Nitridation was performed on a c-plane sapphire substrate at room temperature for 30 min. To investigate the chemical changes caused by plasma nitridation, XPS peaks for Al 2p, N 1s, O 1s, and C 1s were monitored. As examples, the curve-fitting results for the spectra with Al 2p and N 1s at an incident angle of 15° for the DPS and RPS are shown in Fig. 3.3. Four different components are identified in the N 1s photoelectron energy regions. The binding energy of 397.3 eV can be assigned to N–Al bonding [11, 12]. The lowest binding energy peak corresponds to the substitution of N atoms in the O sublattice [(N)O], and the highest binding energy peak to the substitution of N molecules in the O sublattice [(N$_2$)O]. The peak at approximately 399 eV has been assigned to the overlap of the N–H and C–N components [13, 14]. Incorporation of H and C was unintentional and may have been due to the transfer process in air

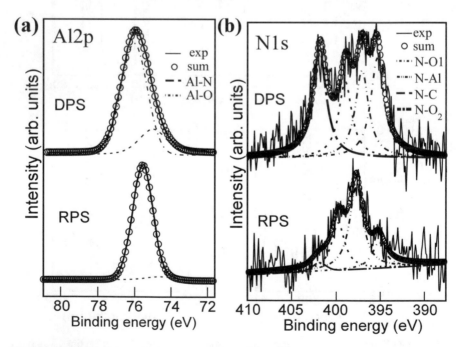

Fig. 3.3 XPS of **a** Al 2p and **b** N 1s at the incident angle of 15° from DPS. Solid lines and open circles indicate spectra and sum-fitted curves, respectively. Dashed lines are fitted curves for each bond

Fig. 3.4 XPS intensity ratios of nitride and oxide of **a** Al 2p and **b** N 1s as a function of measurement angle. N–O indicates the total intensity of (N)O and (N$_2$)O. Circles and squares indicate the direct and remote positions, respectively

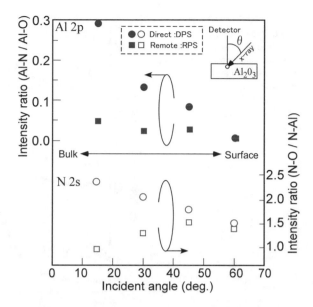

ambient. This peak was observed for an untreated sapphire substrate. On the other hand, in the Al 2p photoelectron energy regions, two different components at 74.4 and 75.7 eV were identified. The peak at 74.4 eV was due to Al–N, whereas the peak at 75.7 eV was identified as an Al 2p peak in Al–O [11, 15]. Angle-resolved XPS measurements revealed the depth profiles of the surface; namely, the angles of 15° and 60° were bulk-sensitive and surface-sensitive, respectively. Figure 3.4 shows the measurement angle dependence of the intensity ratios of Al–N/Al–O and N–O/N–Al. N–O indicated the total intensity of (N)O and (N$_2$)O. The Al–N/Al–O intensity ratio of the DPS was much greater than that of the RPS at a measurement angle of 15°. This means that the DPS had many more Al–N bonds than the RPS. In the near-surface regions of the RPS, the spectral area of the N–O component was greater than that of the N–Al component, whereas in the bulk regions, the N–O spectral area was smaller. The DPS showed a tendency opposite to that of the RPS. The DPS had many more N–O bonds in bulk regions than the RPS. Additionally, a clear difference also appeared in the spectrum of oxygen, as shown in Fig. 3.5. In the O 1s photoelectron energy regions for the DPS, three different components were identified. The peaks at energies of approximately 529.7 and 528.5 eV corresponded to O–Al and O–N, respectively [11, 16], and the other peak corresponded to the surface adsorption layer. The DPS included an O–N bond, although the RPS and the sapphire substrate did not show an O–N peak. The intensities and full widths at half maximum (FWHMs) of the peaks for the RPS were almost the same as those for the sapphire substrate. Losurdo et al. reported on the nitridation mechanism of a c-plane sapphire substrate by remote rf plasmas and described the formation of AlN [17]. This report showed that O–N bonds were an indication of the formation of AlN. The chemical reactions at the sapphire surface involved AlN formation and NO formation

Fig. 3.5 XPS of O 1s from
a DPS, **b** RPS, and
c c-sapphire substrate. Solid
lines and open circles
indicate spectra and
sum-fitted curves,
respectively. Dashed lines
are fitted curves for each
bond

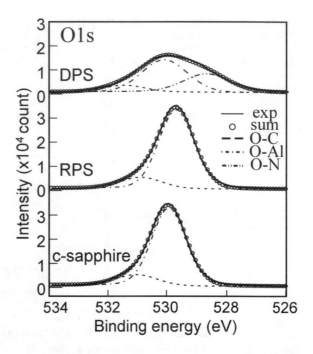

at O sites. As the nitridation of the sapphire surface proceeded, N–O-related XPS peaks increased in intensity. The RPS supported the adsorption of the nitrogen on the surface, and the DPS supported the formation of AlN. A slight peak shift for the RPS was seen at low energies; we consider this to be due to a change in electric charge owing to nitrogen adsorption on the surface. Although the remote plasma formed only surface nitrogen terminations on the sapphire substrate, the direct plasma appeared to form AlN. The nitrogen plasma generated by the AP system has the capability of nitriding a sapphire surface even at room temperature while maintaining a flat surface. This plasma has significant potential for employment in the nitridation of oxide materials at high nitrogen partial pressure and low temperature.

3.3 Near-Atmospheric-Pressure Nitrogen Plasma Passivation

3.3.1 Nitridation of ZnO Surface

Figure 3.6a shows the N 1s spectrum from a near-atmospheric-pressure nitrogen-plasma-treated ZnO substrate (denoted as NAP-ZnO) at the take-off angle (TOA) of 30°. For NAP-ZnO, two different components are identified typically. According to a previous report, the binding energy of 400.1 eV can be assigned to the N_2–O

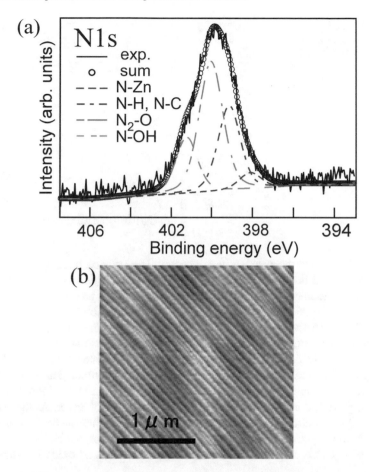

Fig. 3.6 a HX-PES spectra of N 1s from NAP-ZnO at the TOA of 30°. Solid lines and open circles correspond to the experimental data and fitted curve, respectively. Dashed lines represent the fitted curves for each bond: N–H, –C, N–O, and N–Zn. The inset shows the measurement setup. **b** AFM images of NAP-ZnO

component [14]. In relation to the nitrogen AP source, the main constituent of the nitrogen plasma is excited N_2, which is mainly responsible for causing the nitridation of the ZnO surface. The peaks at approximately 399.2 and 401.3 eV have been assigned to the overlap of the N–H and N–C components and N–OH bonding states [14], respectively, the incorporation of which was unintentional and may have been due to the transfer process in ambient atmospheres. A weak peak at 398.0 eV was assigned to the N–Zn bond [18]. These results suggested that the NAP can nitride the ZnO surface even at room temperature, although conventional RF or ECR nitrogen plasma sources require a heating process. Figure 3.6b shows an AFM image of NAP-ZnO, which showed the same step and terrace structure as that of UT-ZnO, suggesting that the plasma treatment did not visibly damage the surface.

Fig. 3.7 HX-PES spectra at
the valence band regions for
UT-ZnO (solid line) and
NAP-ZnO (dashed line) at
the TOA of 86°. The
Fermi-level corresponds to
0 eV

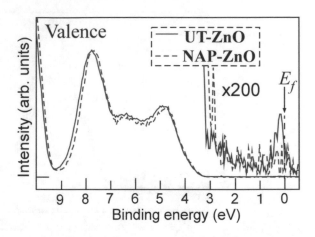

Figure 3.7 shows the valence band spectra. The valence band spectrum of the ZnO substrate was dominated by the O 2p valence band, which had an onset energy, as defined by the linear extrapolation of the valence band maximum (VBM). After nitridation, the VBM shifted to the lower binding energy by 0.2 eV. Note that the core spectra of Zn and O showed same direction shifts as the VBM, indicating that the shifts did not result from the formation of additional bonding and/or defect states but as a change in band filling. Consequently, the Fermi-level moved slightly to the midgap. The onset at the Fermi energy for UT-ZnO contained a weak but well-defined feature associated with occupied conduction band states including donor states, suggesting the existence of the surface accumulation layer. At the VBM of NAP-ZnO, the conduction band feature was reduced to less than the background noise level, providing evidence that the surface accumulation layer, which was considered as a possible cause of a low SBH with a high IF of a metal/oxide interface, was eliminated by NAP treatment.

3.3.2 Improvement of Metal/ZnO Interface

3.3.2.1 Electrical Properties

Figure 3.8 shows the I–V characteristics of the Pt contacts on UT-ZnO and NAP-ZnO. SBH (Φ_B) and IF (n) were determined using Eqs. (2.5) and (2.6). The SBHs of UT- and NAP-ZnO were 0.79 eV ($n = 1.40$) and 0.87 eV ($n = 1.12$), respectively, indicating the reduction of the recombination current. These suggest the improvement of the electrical properties after NAP treatment.

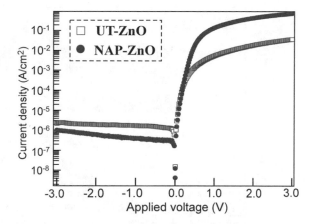

Fig. 3.8 I–V characteristics of UT- (squares) and NAP-ZnO (circles)

3.3.2.2 Chemical Bonding State

Figure 3.9a shows HX-PES spectra of the Zn $2p_{3/2}$ core levels at the TOA $= 30°$ from the Pt/NAP- and UT-ZnO interfaces. The peaks at 1021.3 and 1022.3 eV correspond to the ZnO substrate [19]. As shown in Fig. 2.16, the additional peaks were confirmed at the lower and higher binding energies: ZnPt, ZnPtO, and ZnPtOx, which are caused by Zn diffusion into the Pt layer. After the nitrogen treatment, the intensity ratio of the ZnPt-related peaks to the ZnO substrate decreased from 29 to 14%. At the bulk region (TOA $= 86°$), the peak shape difference between UT- and NAP-ZnO was pronounced, as shown in Fig. 3.9b. The Zn $2p_{3/2}$ core-level spectrum of NAP-ZnO was narrowed. In contrast, the spectra of UT-ZnO still had clear peaks corresponding to Zn in the Pt layer. These results suggest that the amount of Zn diffusion into the Pt layer was much greater than that in NAP-ZnO. The O 1s spectra were overlapped with the high binding energy tail of the Pt $4p_{3/2}$ states, as shown in Fig. 3.9b. By comparing the TOAs of 30 (not shown) and 86°, the Zn–O (530.5 eV) and Pt–O (532.5 eV) bonding states were confirmed [20, 21]. The clear change in the Pt–O bonding state indicated that the direct oxidization of the Pt contact was not observed owing to the weak intensity. However, for the Zn–O bonding state, the peak intensity of UT-ZnO was greater than that of NAP-ZnO by approximately 10% at the TOA $= 86°$, indicating that in contrast to Pt/NAP-ZnO, some of the oxygen of Pt/UT-ZnO existed at the shallower region. By combining the results of Zn $2p_{3/2}$, it is determined that the oxygen was delivered into the Pt layer, since Zn is known as a deoxidant in alloys [22].

3.3.2.3 Interfacial Reaction

To verify the thicknesses and densities of the intermediate layers, X-ray reflectivity (XRR) measurements were performed (Bruker AXS, D8 Discover Super Speed). XRR spectra were analyzed by fitting to extract a simulated reflectivity curve from

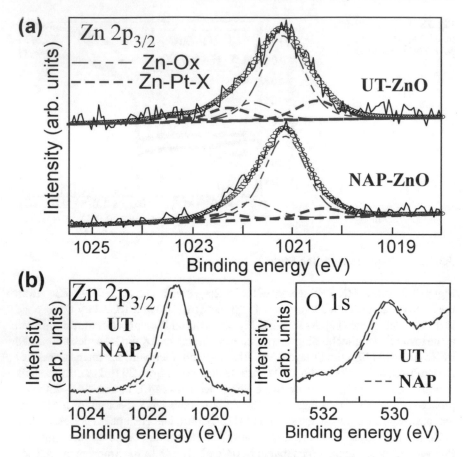

Fig. 3.9 a HX-PES spectra of Zn $2p_{3/2}$ from UT- and NAP-ZnO at the TOA of 30°. Solid lines and open circles correspond to the experimental data and fitted curve, respectively. Dashed lines represent the fitted curves for each bond: Zn-Ox and Zn-Pt-X (Pt-related bonding state). **b** HX-PES spectra of Zn $2p_{3/2}$ and O1s from UT- and NAP-ZnO at the TOA of 86°

the experimental curve using the XRR analysis software DIFFRACplus LEPTOS (Bruker AXS) [23]. A genetic algorithm was used to minimize the logarithm of the difference in the absolute intensity between the simulated and experimental curves, as the model parameters were adjusted using a computer.

After the plasma treatment, the XRR oscillation period of NAP-ZnO became longer than that of UT-ZnO (Fig. 3.10a), suggesting that the interface of Pt/NAP-ZnO had a thinner intermediate layer than that of Pt/UT-ZnO. Figure 3.10 shows the simulated film densities of the intermediate layers of UT-ZnO and NAP-ZnO relative to their respective distances from the ZnO surface. At the interface, the density of the interchanged layer of UT-ZnO was smaller than that of NAP-ZnO by approximately 30%. The estimated roughness of the interface between Pt and ZnO was approximately 0.2 nm, which corresponds to the RMS value of the AFM

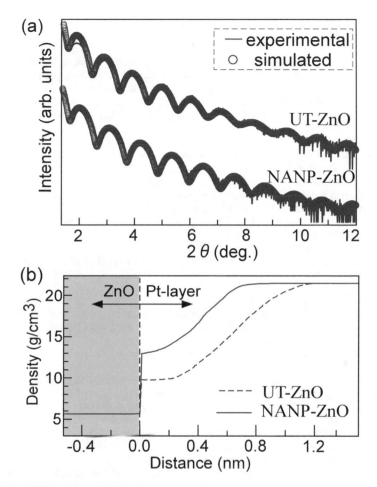

Fig. 3.10 a XRR spectra of UT- and NAP-ZnO. Open circles and solid lines correspond to the experimental data and simulated curves, respectively. **b** Simulated interface structures of UT- and NAP-ZnO

image shown in Fig. 3.6. Consequently, effective interchanged layers for UT- and NAP-ZnO were 0.5 ± 0.1 and 1.0 nm ± 0.1, respectively. These results suggest that the nitrogen passivation also inhibited the diffusion of Zn and O.

3.3.2.4 Summary

We have investigated the effect of NAP treatment on the Pt/ZnO interface. NAP-ZnO showed a significant passivation effect, achieving Schottky contacts of higher quality than those on UT-ZnO. Nitridation of the ZnO surface reduced the formation of a surface accumulation layer and the occurrence of Zn diffusion at the interface between Pt and ZnO.

References

1. Nagata T, Volk J, Yamashita Y, Yoshikawa H, Haemori M, Hayakawa R, Yoshitake M, Ueda S, Kobayashi K, Chikyow T (2009) Interface structure and the chemical states of Pt film on polar-ZnO single crystal. Appl Phys Lett 94:221904. https://doi.org/10.1063/1.3149701
2. Saito M, Wagner T, Richter G, Ruhle M (2009) High-resolution TEM investigation of structure and composition of polar Pd/ZnO interfaces. Phys Rev B 80:134110. https://doi.org/10.1103/PhysRevB.80.134110
3. Allen MW, Swartz CH, Myers TH, Veal TD, McConville CF, Durbin SM (2010) Bulk transport measurements in ZnO: the effect of surface electron layers. Phys Rev B 81:075211. https://doi.org/10.1103/PhysRevB.81.075211
4. Ohashi N, Adachi Y, Osawa T, Matsumoto K, Sakaguchi I, Haneda H, Ueda S, Yoshikawa H, Kobayashi K (2009) Polarity-dependent photoemission spectra of wurtzite-type zinc oxide. Appl Phys Lett 94:122102. https://doi.org/10.1063/1.3103271
5. Nagata T, Bierwagen O, White ME, Tsai MY, Speck JS (2010) Study of the Au Schottky contact formation on oxygen plasma treated n-type SnO_2 (101) thin films. J Appl Phys 107:033707. https://doi.org/10.1063/1.3298467
6. Murarka SP, Chang CC, Adams AC (1979) Thermal nitridation of silicon in ammonia gas: composition and oxidation resistance of the resulting films. J Electrochem Soc 126:996–1003. https://doi.org/10.1149/1.2129223
7. Yuasa M, Yara T (1999) U.S. Patent 5968377
8. Nagata T, Haemori M, Anzai J, Uehara T, Chikyow T (2009) Surface nitridation of c-plane sapphire substrate by near-atmospheric nitrogen plasma. Jpn J Appl Phys 48:040206. https://doi.org/10.1143/JJAP.48.040206
9. Pearse PWB, Gaydon AG (1984) The identification of molecular spectra, 4th edn. Chapman and Hall, New York
10. Powell RC, Lee NE, Kim YW, Greene JE (1993) Heteroepitaxial wurtzite and zinc-blende structure GaN grown by reactive-ion molecular-beam epitaxy: growth kinetics, microstructure, and properties. J Appl Phys 73:189. https://doi.org/10.1063/1.353882
11. Liao HM, Sodhi RNS, Coyle TW (1993) Surface composition of AlN powders studied by x-ray photoelectron spectroscopy and bremsstrahlung-excited Auger electron spectroscopy. J Vac Sci Technol A 11:2681–2686. https://doi.org/10.1116/1.578626
12. Gautier M, Duraud JP, Gressus CL (1987) Electronic structure of an AlN film produced by ion implantation, studied by electron spectroscopy. J Appl Phys 61:574. https://doi.org/10.1063/1.338207
13. Ma JM, Liu YC, Mu R, Zhang JY, Lu YM, Shen DZ, Fan XW (2004) Method of control of nitrogen content in ZnO films: structural and photoluminescence properties. J Vac Sci Technol B 22:94. https://doi.org/10.1116/1.1641057
14. Futsuhara M, Yoshioka K, Takai O (1998) Optical properties of zinc oxynitride thin films. Thin Solid Films 317:322–325. https://doi.org/10.1016/S0040-6090(97)00646-9
15. Heinlein C, Grepstad J, Reichert H, Averbeck R (1997) Plasma preconditioning of sapphire substrate for GaN epitaxy. Mater Sci Eng B 43:253–257. https://doi.org/10.1016/S0921-5107(96)01878-8
16. Wang CC, Chiu MC, Shiao MH, Shieu FS (2004) Characterization of AlN thin films prepared by unbalanced magnetron sputtering. J Electrochem Soc 151:F252–F256. https://doi.org/10.1149/1.1790531
17. Losurdo M, Capezzuto P, Bruno G (2000) Plasma cleaning and nitridation of sapphire (α-Al_2O_3) surfaces: new evidence from in situ real time ellipsometry. J Appl Phys 88:2138. https://doi.org/10.1063/1.1305926
18. Bian JM, Li XM, Gao XD, Yu WD, Chen LD (2004) Deposition and electrical properties of N-In codoped p-type ZnO films by ultrasonic spray pyrolysis. Appl Phys Left 84:541. https://doi.org/10.1063/1.1644331

19. Islam MN, Ghosh TB, Chopra KL, Acharya HN (1996) XPS and X-ray diffraction studies of aluminum-doped zinc oxide transparent conducting films. Thin Solid Films 280:20–25. https://doi.org/10.1016/0040-6090(95)08239-5

20. Bancroft GM, Adams I, Coatsworth LL, Bennewitz CD, Brown JD, Westwood W (1975) ESCA study of sputtered platinum films. Anal Chem 47:586–588. https://doi.org/10.1021/ac60353a050

21. Nagata T, Haemori M, Yamashita Y, Iwashita Y, Yoshikawa H, Kobayashi K, Chikyow T (2010) Oxygen migration at Pt/HfO$_2$/Pt interface under bias operation. Appl Phys Lett 97:082902. https://doi.org/10.1063/1.3483756

22. Schumacher EE, Ellis WC (1932) The deoxidation of copper with the metallic deoxidizers, calcium, zinc, beryllium, barium, strontium and lithium. Trans Electrochem Soc 61:91–100. https://doi.org/10.1149/1.3498000

23. Feranchuk ID, Feranchuk SI, Ulanenkov AP (2007) Self-consistent approach to x-ray reflection from rough surfaces. Phys Rev B 75:085414. https://doi.org/10.1103/PhysRevB.75.085414

Chapter 4
Bias-Induced Interfacial Redox Reaction in Oxide-Based Resistive Random-Access Memory Structure

4.1 Introduction

Resistive switching phenomena have attracted a great deal of attention owing to their high potential for next-generation nonvolatile memory applications such as resistive random-access memory (ReRAM), which is an alternative to well-established practical memories such as static random-access memory (SRAM) and dynamic random-access memory (DRAM) [1, 2]. Many material combinations and structures have been proposed for ReRAM application. A typical resistive switching model is based on a thermal effect initiated by a voltage-induced partial dielectric breakdown that forms a discharge filament modified by Joule heating [3, 4]. The intrinsic material properties also induce changes in resistance. For example, the insulator–metal transition (IMT) in perovskite oxides such as $(Pr,Ca)MnO_3$ [5–7] and $SrTiO_3$:Cr [8] is induced by electronic charge injection operations such as doping.

We focus on a solid electrolyte based on the nanoionics model [9], whose change in resistance was originally demonstrated using chalcogenide materials such as As_2S [10, 11], Cu_2S [12], and GeAs [13]. The fundamental unit cell of this kind of switching device has a simple capacitor-like structure, which is composed of a solid electrolyte sandwiched between two conductive electrodes. In fact, the anode and the cathode electrode materials are generally Ag or Cu, and Pt, respectively. The noteworthy electrical characteristics of these devices are bipolar behavior controlled by the polarity of the applied voltage and their two different resistance states: a high-resistance state (HRS) and a low-resistance state (LRS). However, in terms of a practical large-scale integration (LSI) process and its application, chalcogenide materials are incompatible with the current LSI process owing to their higher vapor pressure and their behavior as contaminants in fabrication equipment. Moreover, the operating voltage is too low (i.e., the ion migration speed is too high) to drive numerous switching devices simultaneously. The matrix materials are also critical as regards controlling the diffusion of Cu. To control the defect density or grain size of the matrix, oxides with ion binding such as ZrO_2 are somewhat better candidates. Taking these factors into account, oxide-based ReRAM based on the nanoionics model has been proposed as a new

© National Institute for Materials Science, Japan 2020
T. Nagata, *Nanoscale Redox Reaction at Metal/Oxide Interface*, NIMS Monographs,
https://doi.org/10.1007/978-4-431-54850-8_4

application for oxides [14–18]. We also investigated suitable electrode and matrix materials for the LSI process and demonstrated the resistance switching behavior of the Cu/HfO$_2$/Pt system [19]. We used HfO$_2$, which is known as an ionic crystal, and a Cu electrode as a mobile ion source owing to its faster diffusion to create filamentary paths in a matrix. HfO$_2$ is currently playing an important role in global LSI trends including as a high-k gate insulator for advanced complementary metal/oxide semiconductor (CMOS) technologies [20, 21]. Moreover, with the miniaturization of integrated circuits, Cu is being used for LSI interconnection wire because its conductivity is superior to that of Al. For this reason, HfO$_2$ and Cu seem to be more suitable for the LSI process than other materials. Therefore, there has been increased interest in the use of high-k dielectric oxides as potential ReRAM materials.

Both the nanoionics model and the vacancy model, which is based on oxygen vacancy nucleation at a metal/oxide interface, have been proposed as the mechanism of the resistance switching behavior of an oxide-based ReRAM with a metal–oxide–metal (MOM) structure [22–24]. Yang et al. indirectly observed that oxygen vacancies are created and drift toward the cathode, forming localized conducting channels on Pt/TiO$_2$/Pt cross-point junctions [22]. In contrast, Sakamoto et al. reported the creation of Cu metal filaments in Ta$_2$O$_5$ via an electrochemical reaction [15]. Although these different mechanisms must result from the use of different material combinations, there has been no direct observation at a metal/oxide interface under bias operation.

As shown in Chap. 2, HX-PES is a powerful tool for investigating the chemical state of the interface of stack structures for nanoelectronic devices without causing degradation. By this method, bias-induced compositional changes around the metal/oxide interface during device operation have been directly observed [25, 26].

Although the formation of filaments on an anode has been reported by several groups [10, 15, 18, 27], there has been no direct observation of metal ionization and diffusion at a metal/oxide interface under bias operation. Here, we used HX-PES with an applied bias to examine the electronic structure at the interface of a Cu/HfO$_2$/Pt (MOM) structure in an operating device. In this chapter, we report on the resistance switching behavior and direct observation of the filament formation process of the Cu/HfO$_2$/Pt structure compared with those for a Pt/HfO$_2$/Pt structure, which is an example of a structure based on the oxygen vacancy model.

4.2 Nanoionic-Type ReRAM Structure

4.2.1 Sample Structure

Figure 4.1a shows the structure of a sample device used in our investigation. Insulating c-plane sapphire (Al$_2$O$_3$) was used as a substrate to eliminate additional factors responsible for electrical conductivity. A HfO$_2$ layer was deposited by pulsed laser deposition with a KrF excimer laser ($\lambda = 248$ nm) on a 100-nm-thick Pt bottom

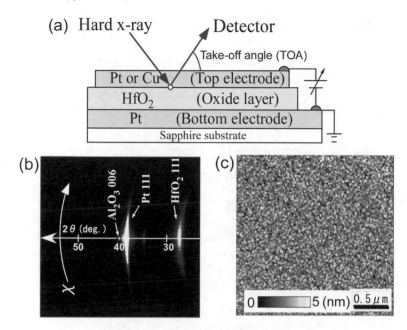

Fig. 4.1 **a** Schematic illustration of measurement setup for HX-PES under bias application. **b** 2D-XRD and **c** AFM images of HfO$_2$/Pt/Al$_2$O$_3$ structure

electrode deposited on the substrate by DC sputtering. During the deposition, the substrate temperature was maintained at 300 °C under an O$_2$ partial pressure of 1 $\times$ 10^{-3} Torr. The crystal structure of the HfO$_2$ thin film was identified by XRD. Figure 4.1b shows the XRD pattern of a HfO$_2$ thin film deposited on Pt/Al$_2$O$_3$. The diffraction peaks of the HfO$_2$ thin film exhibit a ring pattern derived from the polycrystalline monoclinic structure. The strong intensity of the (111) peak means that the HfO$_2$ grains are preferentially oriented in the [111] direction. The surface morphology of the HfO$_2$ film was characterized by AFM, as shown in Fig. 4.1c. The obtained root-mean-square (RMS) roughness was 0.6 nm. The surface morphology also indicated that the HfO$_2$ thin film consisted of fine 20 nm grains. For the HX-PES measurement under bias application, gold wires fixed to the top and bottom electrodes with Pt paste were connected to electrodes on the sample stage. The thickness of the top and bottom electrodes (Cu or Pt) was 10 nm. A bias voltage was applied to the sample with a DC voltage current source/monitor. HX-PES measurements were performed at take-off angles (TOAs) of 85°, 40°, and 20° to obtain rough depth profiles; the TOA of 20° was surface/interface-sensitive, whereas the TOA of 85° was bulk-sensitive.

4.2.2 Electrical Properties of Cu/HfO₂/Pt Structure

I–V measurements were performed on the sample with a semiconductor parameter
analyzer at room temperature. A positive bias was applied to the top Cu electrode
and the Pt electrode was grounded. Figure 4.2a shows the I–V characteristic of the
Cu/HfO₂/Pt structure. The resistance changed when an applied voltage was swept

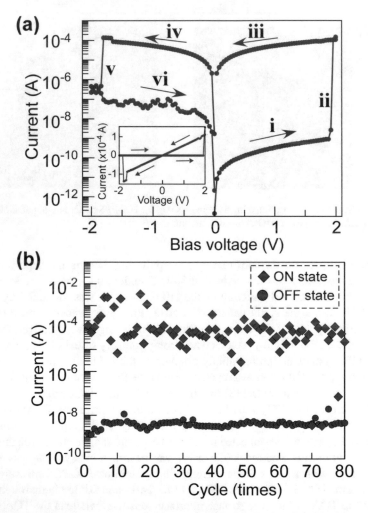

Fig. 4.2 **a** Typical I–V characteristic of Cu/HfO₂/Pt structure at room temperature. Arrows indicate
the voltage sweep direction (i → ii → iii → iv → v→vi). i and vi correspond to the high-resistance
state (HRS), iii and iv correspond to the low-resistance state (LRS), and ii and v are the turn-on and
turn-off voltages, respectively. The inset shows a plot on linear axes. **b** Pulsed I–V characteristic
of Cu/HfO₂/Pt structure at room temperature. The pulse width, input positive voltage (set voltage),
and negative voltage (reset voltage) are 0.1 s, 4.25 V, and −2.25 V, respectively

from a positive bias to a negative bias, indicating that there are two different resistive states at positive and negative voltages; one is the HRS (turn-off) and the other is the LRS (turn-on). An abrupt change in resistance from the HRS to the LRS was induced by increasing the applied voltage to 2.0 V, at which an increase in current of more than four orders of magnitude was observed. When the voltage was swept continuously from a positive to negative voltage, there was a rapid drop in current from the LRS to the HRS at −1.8 V, where a decrease in current of approximately three orders of magnitude was observed. The inset of Fig. 4.2a shows the I–V characteristic plotted on linear axes, which indicated good linearity. Note that the oxygen vacancy model based on oxygen migration at the metal/oxide interface shows nonlinear I–V curves during the formation of oxygen vacancies [28, 29], meaning that, in the case of the Cu/HfO$_2$ interface, the majority of conducting paths in the oxide should be metallic.

We obtained stable switching in each device under low-voltage operation. The transition voltage from the HRS to the LRS was similar for most of the devices. A pulsed I–V measurement was also performed, the results of which are shown in Fig. 4.2b. Stable switching behavior with an average on/off ratio of 10^4 was observed with an input pulse width of 0.1 s.

The turn-on voltage was dependent on the HfO$_2$ film thickness, as shown in Fig. 4.3. As the HfO$_2$ thickness decreased, the turn-on voltage decreased. Furthermore, for an amorphous-phase HfO$_2$ layer, the turn-on voltage increased three-fold. The AFM observation and XRD characterization revealed that the deposited HfO$_2$ thin films had many grain boundaries with a columnar structure. Note that conductive paths form more easily in polycrystalline films with a columnar structure than in amorphous films because there are many possible diffusion paths in polycrystalline films.

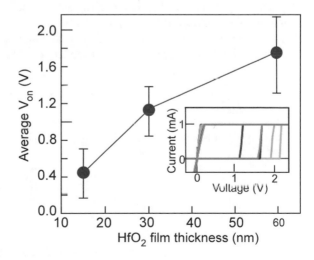

Fig. 4.3 HfO$_2$ film thickness dependence of average turn-on voltage. The inset shows an example of the I–V property of a Cu/60 nm HfO$_2$/Pt structure in the forward bias region

4.3 HX-PES Measurements Under Bias Application

4.3.1 Cu/HfO₂ Interface

During the HX-PES measurement of the Cu/HfO$_2$/Pt structure, the current was measured, as shown in Fig. 4.4. The I–V property exhibited resistance switching behavior from the HRS to the LRS at a bias of 1.3 V, at which an increase in current of three orders of magnitude occurred. To investigate three different resistance states, we performed angle-resolved HX-PES measurements at biases of 0 V (initial stage), 0.4 V (HRS), and 1.3 V (LRS). Above 1.3 V, the high current made the vacuum level of the system unsuitable for obtaining measurements.

Figure 4.5a shows the Cu $2p_{3/2}$ spectrum. In the Cu $2p_{3/2}$ spectrum, the peak at a binding energy of 932.6 eV corresponds to metallic Cu [30, 31]. The most common oxides of Cu are CuO and Cu$_2$O. According to previous studies, the Cu $2p_{3/2}$ core level of CuO is about 1.0 eV higher than that of metallic Cu and has satellite peaks around 941 eV, which were not observed in the Cu $2p_{3/2}$ spectrum. As the TOA decreased, the Cu 2p spectrum (Fig. 4.5b) had a stronger core-level peak at 932.4 eV, which was attributed to Cu$_2$O, indicating that the Cu electrode had Cu$_2$O layers at the surface and interface [30, 31]. Therefore, the spectrum was fitted by those for Cu metal and Cu$_2$O, as shown in Fig. 4.6a. According to a previous report, a Cu film stored under atmospheric conditions for a few days formed a Cu$_2$O layer with a thickness of approximately 0.8–1.0 nm [32, 33]. Additionally, a 5 nm Pt/5 nm Cu/HfO$_2$ structure, which was deposited via an in situ process in a high-vacuum system to eliminate the surface oxidization of the Cu layer (making it an unsuitable sample for studying O 1s spectra owing to the overlapping spectra of the O 1s and Pt $4p_{3/2}$ core levels), was investigated by HX-PES to estimate the thickness of the interfacial oxidized layer between Cu and HfO$_2$. The investigation revealed that a 0.9 ± 0.1 nm CuO$_2$ layer is formed at the interface of the as-deposited 5 nm Pt/5 nm Cu/HfO$_2$ structure. As a consequence, the Cu/HfO$_2$ interface is expected

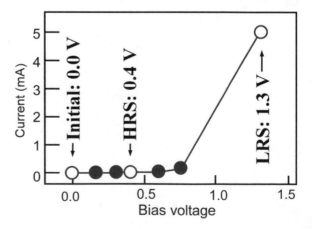

Fig. 4.4 I–V property of Cu/HfO$_2$/Pt structure during HX-PES measurement. Open circles are points measured by HX-PES

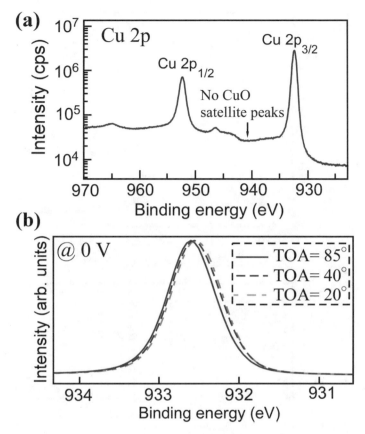

Fig. 4.5 **a** Cu 2p spectrum obtained by HX-PES for the Cu/HfO$_2$/Pt structure under a bias of 0 V at a TOA of 0°. **b** TOA dependence of Cu 2p$_{3/2}$ spectrum obtained by HX-PES under a bias of 0 V

to have an approximately 1.8-nm-thick CuO$_2$ layer. At the surface, the O 1s core level attributed to the adsorbed hydro-oxygen and the C 1s core level exhibited no applied bias dependence (Fig. 4.6b inset), indicating that the application of a bias did not affect the surface state. Furthermore, the inset in Fig. 4.6a shows a difference peak, which is the Cu 2p$_{3/2}$ peak at 0 V minus that at 1.3 V. As the bias increased, the intensity of the Cu$_2$O peak decreased, as shown in Fig. 4.6b, suggesting the reduction of the Cu$_2$O layer at the interface. To examine the volume of the reduced Cu$_2$O layer, we estimated the thickness of the Cu$_2$O layer (d) using the following equation [34]:

$$d = \lambda_{Cu_2O} \cdot \sin\theta \cdot \ln\left[\frac{\delta_{Cu} \cdot M_{Cu_2O} \cdot \lambda_{Cu}}{\delta_{Cu_2O} \cdot M_{Cu} \cdot \lambda_{Cu_2O}} \cdot \frac{I_{Cu_2O}}{I_{Cu}} + 1\right], \qquad (4.1)$$

where λ is the IMFP, δ is the density, M is the molar weight, I is the HX-PES intensity for Cu and Cu$_2$O, and θ is the TOA. The surface Cu$_2$O layer was estimated to be 0.9 nm thick. Using the results obtained at TOAs of 85° and 20°, the estimated

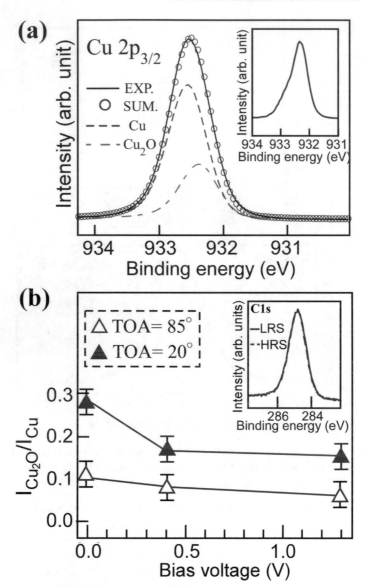

Fig. 4.6 **a** Cu $2p_{3/2}$ spectrum obtained by HX-PES for the Cu/HfO$_2$/Pt structure under a bias of 0 V at a TOA of 20°. The inset shows a difference peak, which is the Cu $2p_{3/2}$ peak at 0 V minus that at 1.3 V. **b** Applied bias dependence of intensity ratio of Cu$_2$O to Cu. The inset shows the C $1s$ spectrum obtained by HX-PES at biases of 0.4 V (HRS) and 1.3 V (HRS) and a TOA of 20°

interfacial Cu_2O thicknesses at voltages of 0 and 1.3 V were approximately 0.9 ± 0.1 and 0.1 ± 0.1 nm, respectively, indicating that the interfacial Cu_2O layer was reduced by the application of bias. Note that no CuO peak was confirmed during the bias application measurements.

In the Hf $3d_{5/2}$ spectrum (Fig. 4.7a), the peak at 1662.0 eV corresponds to HfO_2 [35]. The intensity of the Hf $3d_{5/2}$ spectrum tail at a low binding energy increased as the bias voltage increased. On the basis of the electronegativity, the tail of the Hf $3d_{5/2}$ spectrum was assigned to the Cu–Hf–O bonding state. The intensity ratio of Cu–Hf–O/Hf–O increased with the bias at both TOAs, as shown in Fig. 4.7b. Furthermore, the increase in the ratio at a TOA of 20° was greater than that at a TOA of 85°. Figure 4.7b also summarizes the intensity ratio of Cu $2p_{3/2}$/Hf $3d_{5/2}$ against the bias voltage. The ratio decreased with increasing bias voltage at both TOAs. Furthermore, after removing the bias, the HX-PES spectrum for the metal maintained the same shape as that at 1.3 V, as shown in Fig. 4.8. These results provide evidence that the Cu_2O at the interface was reduced and diffused into the HfO_2 layer.

In the O 1s spectrum, four oxygen peaks were identified, as shown in Fig. 4.9a. The highest binding energy peak at 532.2 eV corresponded to O–OH components containing hydrogen adsorbed on the surface. The peaks at 530.7 and 531.4 eV are attributable to the O–Hf and O–Cu(+1) bonding states, respectively [30, 31, 36]. The lowest binding energy peak at 529.8 eV was assigned to O–Hf–Cu. The behavior of the O–Cu(+1) peak, which exhibited a reduction in the intensity relative to the total intensity of O 1s with increasing bias voltage, as shown in Fig. 4.9b (solid circles), was consistent with that of the Cu $2p_{3/2}$ peak. Furthermore, the O–Hf–Cu peak exhibited a monotonic increase with increasing bias voltage (Fig. 4.9b solid squares). These results are consistent with the changing behavior of the Cu $2p_{3/2}$ and Hf $3d_{5/2}$ spectra with the bias voltage. To determine whether oxygen vacancies formed in the HfO_2 layer upon bias application, the applied bias dependence of the intensity ratio of O–Hf (open circles) to the total intensity of O 1s is also plotted in Fig. 4.9b. No applied bias dependence was observed, meaning that oxygen vacancies did not form at the interface.

4.3.2   Pt/HfO₂ Interface

To investigate the effect of the electrode materials, we fabricated a $Pt/HfO_2/Pt$ structure with the same HfO_2 thickness in the same manner as described earlier. When we used Pt as the top electrode, the I–V curve did not show a hysteresis loop until a voltage sweep had been applied a few tens of times. The different switching behavior of the $Cu/HfO_2/Pt$ and $Pt/HfO_2/Pt$ devices might result from their structural differences.

In HX-PES measurements with an applied bias voltage, the I–V property behaved nonlinearly (Fig. 4.10a). At an applied voltage of 150 mV, the current was 15 mA. Above 150 mV, the high current made the vacuum level of the system unsuitable for carrying out measurements. However, at an applied voltage of 150 mV, a clear difference was observed between the O–Pt and O–Hf states at a TOA of 20° (interface-

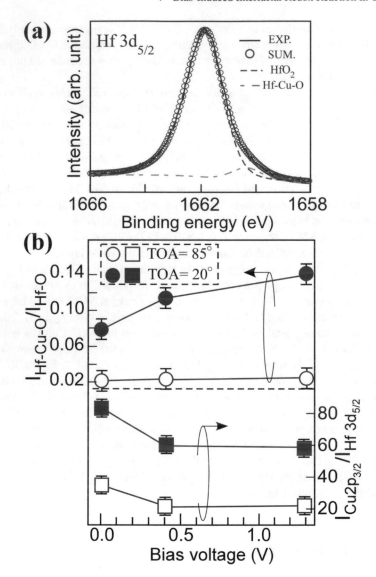

Fig. 4.7 **a** Hf 3d$_{5/2}$ spectrum obtained by HX-PES at 1.3 V and a TOA of 20°. The solid line shows the measured spectrum and open circles show the sum-fitted curve. The dashed lines are fitted curves for each bond: Cu, Cu$_2$O, HfO$_2$, and Hf–Cu–O. **b** Applied bias dependences of intensity ratios of Hf–Cu–O to Hf–O (circles) and Cu 2p$_{3/2}$ to Hf 3d$_{5/2}$ (squares)

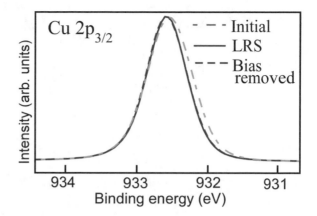

Fig. 4.8 Cu $2p_{3/2}$ spectrum obtained by HX-PES at initial state (0 V), at LRS (1.3 V), and after bias removal at a TOA of 20°

and/or surface-sensitive condition) (Fig. 4.10b). In the O 1s spectrum, because the high binding energy tail of the Pt $4p_{3/2}$ states overlapped with the low-intensity O 1s states, the background could not be removed correctly. To obtain the initial compositional depth profile, HX-PES measurements were performed at several TOAs and at 0 V (Fig. 4.10b). At a TOA of 85°, the peak at around 530.5 eV is attributable to O–Hf bonding. As the TOA increased, the intensity of the O–Hf peak decreased and a peak developed at around 532.8 eV, which is attributed to O–Pt bonding [37, 38], indicating that there was PtO at the interface and/or surface. Note that when the thickness of the Pt layer increased, these peak intensities decreased. Therefore, these states are predominantly formed at the interface. At 0 V, both the –Hf and O–Pt peaks were small. When a forward bias voltage was applied, the O–Pt peak intensity increased, whereas the O–Hf peak intensity decreased, indicating that the Pt interface layer was oxidized and that oxygen vacancies were formed in the HfO_2 layer. Under a reverse bias, the O–Pt peak became more pronounced than under a forward bias, suggesting that the PtO layer formed by the forward bias remained, whereas the reverse bias formed an additional PtO layer at the interface.

Figure 4.11a shows the Pt 4f spectrum at a TOA of 20° and a bias of −150 mV. The Pt 4f HX-PES spectrum was dependent on the applied voltage, and the intensity of the spectrum tail at a high binding energy increased with the applied voltage. In the Pt 4f spectrum, the peaks at binding energies of 71.0 and 74.3 eV correspond to metallic Pt, and the small peaks at 72.2 and 75.5 eV are attributed to PtO. As the forward bias increased, the Pt–O intensity increased (Fig. 4.11b). This indicates that the forward bias voltage drew oxygen toward the anode and induced Pt oxidization, although the oxidization energy of Pt is higher than that of Hf. After the forward bias was removed, the Pt–O intensity decreased slightly but did not revert to the initial intensity ratio at zero bias, meaning that Pt–O bonding remained at the interface after the application of bias.

In contrast, the Hf $3d_{5/2}$ HX-PES spectrum did not show any dependence on the application of forward bias. In the Hf $3d_{5/2}$ spectrum, under a reverse bias (Fig. 4.12a),

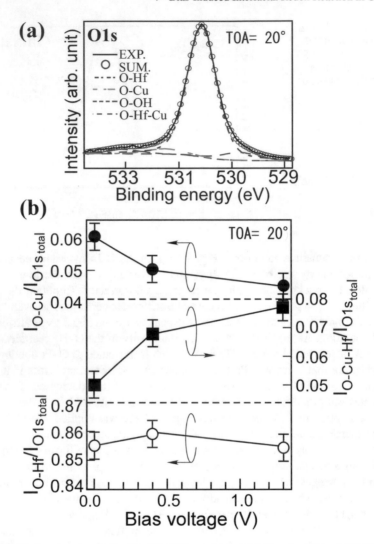

Fig. 4.9 **a** O 1s spectrum obtained by HX-PES at a bias of 1.3 V and a TOA of 20°. The solid line shows the measured spectrum and open circles show the sum-fitted curve. The dashed lines are fitted curves for each bond: O–Hf, O–Cu(+1), O---OH, and O–Hf–Cu. **b** Applied bias dependences of intensity ratios of O–Cu (solid circles), O–Hf–Cu (solid squares), and O-Hf (open circles) to total intensity of O 1s at a TOA of 20°

an additional peak appeared at a lower binding energy, which is assigned to the Hf–Pt bond upon considering the difference in the electronegativity of each atom and referring to previous studies [35]. The Hf–Pt intensity, which showed a monotonic increase with increasing reverse bias, is shown in Fig. 4.12b. This indicates that the reverse bias drew Hf toward the negatively biased anode and induced Hf–Pt bond formation. To examine the bias-induced variation in the Hf density at the interface,

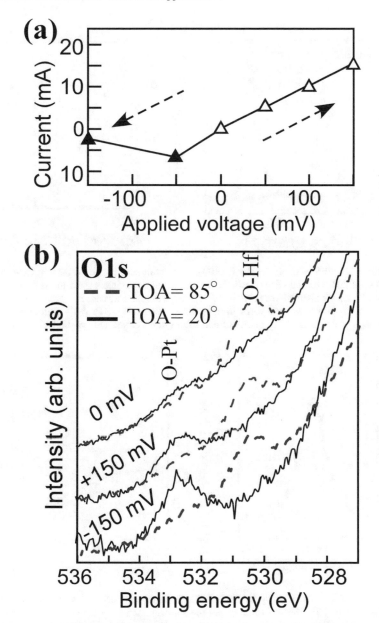

Fig. 4.10 **a** I–V property of Cu/HfO$_2$/Pt structure obtained during HX-PES measurements. **b** HX-PES of O 1s as a function of bias at TOAs of 85° (solid lines) and 20° (dashed lines)

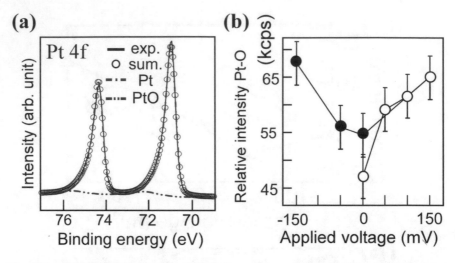

Fig. 4.11 **a** HX-PES of Pt 4f for the Pt/HfO$_2$/Pt structure at a bias of −150 mV (reverse bias) and a TOA of 20°. The solid line shows the spectrum and open circles show the sum-fitted curve. Dot-dashed lines are fitted curves for Pt and PtO bond. **b** Applied bias dependence of Pt 4f intensity of PtO (circles). All values were normalized by the Pt 4f intensity of Pt at 0 V (initial state). Open symbols represent forward-bias conditions and solid symbols represent reverse-bias conditions

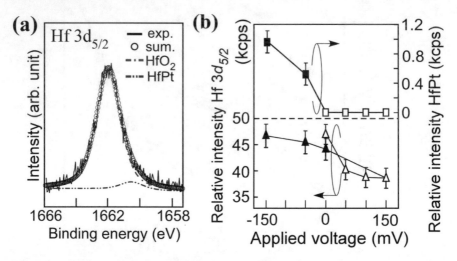

Fig. 4.12 **a** HX-PES of Hf 3d$_{5/2}$ for the Pt/HfO$_2$/Pt structure under a bias of −150 mV (reverse bias) at a TOA of 20°. The solid line shows the spectrum and open circles show the sum-fitted curve. Dot-dashed lines are fitted curves for HfO$_2$ and HfPt bonds. **b** Applied bias dependences of Hf 3d$_{5/2}$ intensity ratios of HfPt (squares) and Hf 3d$_{5/2}$ (triangles). All values were normalized by the Pt 4f intensity of Pt at 0 V (initial state). Open symbols represent forward-bias conditions and solid symbols represent reverse-bias conditions

the Hf $3d_{5/2}$/Pt $4f_{7/2}$ elemental intensity ratio at a TOA of $20°$ was plotted as a function of the bias voltage (Fig. 4.12b). The ratio decreases with increasing forward bias and increases with increasing negative reverse bias. At -150 mV, the ratio reverts to the initial intensity ratio. These results indicate that the forward and reverse biases drew Hf toward the cathode and the negatively biased anode, respectively. Furthermore, the Hf drawn toward the anode induced the oxidization of Pt as indicated by the increase in the PtO intensity in Fig. 4.12b. This suggests that the diffusion of Hf into the Pt anode produced Pt–Hf mixed oxide and stabilized the oxygen atoms in the negatively biased anode. According to a previous study, the rare earth and transition metal affected the valence of Pt and induced Pt–rare earth–O bonding [39].

4.4 Filament Formation Process in Cu/HfO$_2$/Pt and Pt/HfO$_2$/Pt ReRAM Structures

Using the results of HX-PES, we revealed the relationship between the interface structures and the chemical states at the Cu/HfO$_2$/Pt structure in different resistance states, and compared it with the relationship for the Pt/HfO$_2$/Pt structure as illustrated in Fig. 4.13. There is a natural oxidized copper (Cu$_2$O) layer at the Cu/HfO$_2$ interface in the initial state (Fig. 4.13a-i). The application of a bias reduces the interfacial Cu$_2$O layer (Fig. 4.13a-ii) because the standard reduction potential of Cu ions (Cu$^+$ + e$^-$ $\rightarrow$ Cu $+0.52$ V) is lower than that of Pt (Pt^{2+} + 2e$^-$ $\rightarrow$ Pt $+1.19$ V) [40]. Furthermore, as regards metal segregation in a film, Cu segregates and diffuses into a Hf-based film more easily than Pt [41]. It is well known that Ag can be a fast mobile ion in oxides similarly to Cu. From previous reports and our results, it seems to be feasible that Cu migrates in HfO$_2$ to form filamentary paths more easily owing to its fast diffusion in HfO$_2$ through the columnar grain boundaries.

This Cu diffusion behavior is considerably different from the Pt/HfO$_2$ interfacial reaction under bias operation (Fig. 4.13b), namely the oxidization of the Pt layer. The application of a forward bias causes oxygen to diffuse into the Pt layer, thus forming PtO (Fig. 4.13b-ii). Note that this oxygen vacancy formation behavior is consistent with the previously reported electrical properties, namely, that the oxygen vacancies, which act as trap centers for charge carriers, cause the deterioration of the metal/HfO$_2$ interface during CMOS applications [42, 43]. Furthermore, the oxygen vacancies in a high electric field form percolating conduction filaments during ReRAM applications.

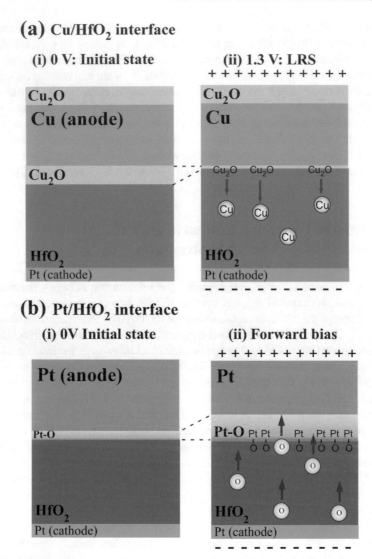

Fig. 4.13 Schematic illustration of **a** Cu/HfO$_2$/Pt structure (i) in initial state and (ii) under bias operation, and **b** Pt/HfO$_2$/Pt structure (i) in initial state and (ii) under bias operation

4.5 Bias-Induced Cu Migration Behavior in Cu/HfO$_2$ ReRAM Structure

The Cu ion migration behavior in the Pt/Cu/HfO$_2$/Pt structure was observed by HX-PES under bias operation. Figure 4.14 shows the I–V characteristics obtained during the HX-PES measurements with a continuously applied bias voltage. The I–V characteristics exhibited resistance switching behavior. The Cu 2p$_{3/2}$, O 1s,

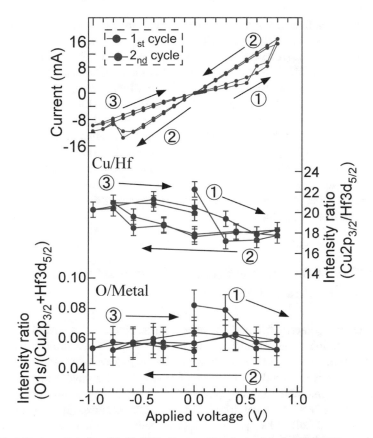

Fig. 4.14 I–V characteristics of Pt/Cu/HfO$_2$/Pt structure obtained during HX-PES measurements, and applied bias dependences of intensity ratios of Cu 2p$_{3/2}$ to Hf 3d$_{5/2}$ and O1s to the sum of Cu 2p$_{3/2}$ and Hf 3d$_{5/2}$. Arrows indicate the voltage sweep direction (1 → 2 → 3)

and Hf 3d$_{5/2}$ spectra had similar structures to those in 3.3.1. After the application of bias, the Cu$_2$O peak intensity was reduced. The application of bias induced the reduction of the unintentionally oxidized Cu$_2$O layer at the Cu/HfO$_2$ interface, which is consistent with the results in 3.3.1. The integration time of the I–V measurement at each voltage during the HX-PES measurements was more than 20 min, which is considerably longer than that of conventional I–V measurements by a source measurement unit (SMU). Consequently, the HX-PES spectra indicated the saturated state after the reaction. In actual devices, the amount of reduction should be less than that measured by HX-PES under bias operation. After switching to the HRS from the LRS, the reoxidization of the Cu layer was not observed, meaning that the bias-induced interfacial reduction/oxidization during the switching is not a reversible reaction.

The Cu diffusion behavior was investigated through the bias voltage dependence of the intensity ratio of Cu 2p$_{3/2}$ to Hf 3d$_{5/2}$, as shown in Fig. 4.14. In the forward bias region, corresponding to switching from the HRS to the LRS, the ratio decreased by $23 \pm 5\%$, suggesting that Cu diffused into the HfO_2 layer. In contrast, in the reverse bias region, corresponding to switching from the LRS to the HRS, the ratio increased by $15 \pm 5\%$. These results imply that the application of bias moves Cu ions in the HfO_2 layer during the switching. After switching to the HRS, the intensity ratio did not return to its initial value. The initial drop of the ratio should correspond to the reduction of the Cu_2O layer. To investigate the correlation between the Cu migration behavior and the oxygen concentration, the intensity ratio of O 1s to the total intensity of Cu $2p_{3/2}$ and Hf $3d_{5/2}$ was also obtained, as shown in Fig. 4.14. After the initial reduction of Cu_2O, the intensity ratio showed no bias voltage dependence, meaning that the oxygen near the interface was stable during the switching. These results suggest that the forming and switching processes at the Cu/HfO_2 interface are not reversible reactions.

4.6 Effect of Bottom Electrode on Interfacial Structure and Switching Behavior

As mentioned in Sect. 4.3.2 and in other reports, electrode metals affect oxygen vacancy formation at the metal/HfO_2 interface [44–46]. According to the oxygen vacancy model, TiN can form conductive filaments more easily than a Pt electrode. A TiN electrode has an oxidized interlayer because the HfO_2 layer has many oxygen vacancies. This interlayer reduces the forming and set voltages and improves retention. Here, the effect of the electrode on the oxygen vacancies is compared in two nanoionics-based ReRAM structures: Cu/HfO_2 ReRAM structures with a Pt bottom electrode and with a TiN bottom electrode.

4.6.1 Electrical Properties of Pt/Cu/HfO$_2$/Pt/Si and Pt/Cu/HfO$_2$/TiN/Si Structures

Figure 4.15 shows the I–V properties of the Pt/Cu/HfO_2/Pt/Si (denoted as Pt-HfO_2) and Pt/Cu/HfO_2/TiN/Si (denoted as TiN-HfO_2) structures. Resistance switching behavior was confirmed for both structures. For conductive filament formation, TiN-HfO_2 required a higher bias voltage (3–4 V) than Pt-HfO_2 (~1.3 V). TiN-HfO_2 also exhibited a nonlinear I–V curve, whereas Pt-HfO_2 exhibited a linear I–V curve. The majority of formation mechanisms of conducting paths processed for Pt-HfO_2 and TiN-HfO_2 are described by the nanoionics and oxygen vacancy models, respectively [8, 47]. Furthermore, memory windows become narrower as the sweep number increases.

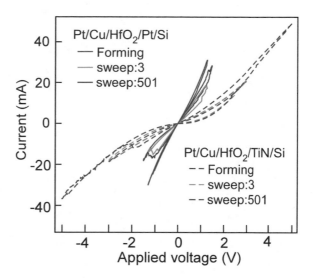

Fig. 4.15 I–V properties of Pt/Cu/HfO$_2$/Pt/Si (solid lines) and Pt/Cu/HfO$_2$/TiN/Si (dashed lines) stack structures

4.6.2 Interfacial Structure Between Cu and HfO$_2$

For Pt-HfO$_2$, as described in 4.3.1, the Cu 2p$_{3/2}$ spectrum consisted of Cu$_2$O (932.9 eV) and Cu metal (932.6 eV) bonding states. The Cu electrode was oxidized at the Cu/HfO$_2$ interface. Furthermore, the Hf 3d$_{5/2}$ peak implied the slight diffusion of Cu into the HfO$_2$ layer.

For TiN-HfO$_2$, the O 1s spectrum at a TOA of 20° had a strongly asymmetric shape, as shown in Fig. 4.16a. This phenomenon has two possible causes. One is the chemical shift corresponding to the chemical reaction. In this case, the O 1s spectrum and the related metal bonding state should shift in opposite directions. The other is band bending behavior near the interface related to Fermi pinning due to the formation of electron states caused by the oxygen vacancies and/or defects, referred to as defective states. In this case, the spectrum of O 1s and the related metal bonding state should shift in the same direction. Consequently, the Cu 2p$_{3/2}$ and Hf 3d$_{5/2}$ core-level spectra should indicate the cause of the above phenomenon. The Cu 2p$_{3/2}$ spectrum shown in Fig. 4.16b, which consists of Cu and Cu$_2$O bonding states, did not clearly exhibit a shift in the binding energy to a higher energy or an additional bonding state consisting of O 1s. In contrast, Hf 3d$_{5/2}$ exhibited an increase in the shoulder peak at a higher binding energy and a peak at around 1663 eV, consistent with the O 1s peak, and no additional bonding state was observed on the lower binding energy side. On the basis of these results, the energy shift of Hf 3d$_{5/2}$ in the same direction as that of O1s and the change in the asymmetric shape with the TOA can be accommodated by band bending and the defective states (Fermi-level pinning). To analyze these spectra, the HfO$_2$ bonding state was investigated by simulating the Fermi-level pinning and band bending using the Common Data Processing System (COMPRO; ver.11) written by Yoshihara and Yoshikawa [48, 49]. The PES spectrum $F(E - E', z_i) = A(z_i) \times S[E - E'(z_i)]$ originating at depth z_i in a detectable region

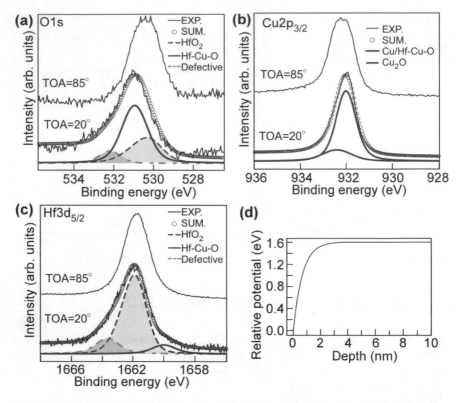

Fig. 4.16 HX-PES of **a** O1s, **b** Cu $2p_{3/2}$, and **c** Hf $3d_{5/2}$ core levels for Pt/Cu/HfO$_2$/TiN/Si stack structure at TOAs of 85° and 20°. Solid lines show measured spectra and open circles show sum-fitted curves. Dashed lines are fitted curves for each bonding state: HfO$_2$, Cu$_2$O, Cu, Hf–Cu–O, and defective HfO$_2$. **d** Energy potential of HfO$_2$ layer at the Cu/HfO$_2$ interface of the Pt/Cu/HfO$_2$/TiN/Si stack structure simulated by COMPRO. The Cu/HfO$_2$ interface is set at a depth of 0 nm

can be simulated by combining a binding energy shift $E'(z_i)$ caused by band bending and the attenuation of intensity $A(z_i) = \exp(-z_i/\lambda\cos\theta)$, where λ is the IMFP and θ is the TOA. After integrating the simulated spectrum $F(E - E', z_i)$ over z_i, we minimize the chi-square value of the difference in the normalized intensity between the simulated and experimental spectra at several TOAs while adjusting the model parameters of the energy distribution using COMPRO. In Fig. 4.16a, c, the spectra corresponding to the band bending and the defective states are, respectively, indicated as green and red dashed lines. Figure 4.16d shows the simulated potential of HfO$_2$. HfO$_2$ clearly exhibits downward band bending, which is the movement of the Fermi level toward the conduction band, from the bulk region to the Cu/HfO$_2$ interface. Additionally, TiN-HfO$_2$ has a thicker unintentionally oxidized Cu$_2$O layer at the interface than Pt-HfO$_2$. To estimate the volume of the interfacial oxidized layer, the thickness of the Cu$_2$O layer was estimated using Eq. (3.1). The estimated interfacial Cu$_2$O thicknesses for TiN-HfO$_2$ and Pt-HfO$_2$ were approximately 3.9 ± 0.2 and $0.6 \pm$

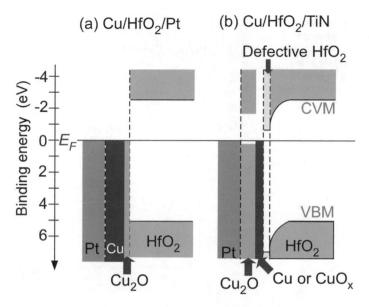

Fig. 4.17 Schematic energy level diagrams for **a** Pt/Cu/HfO$_2$/Pt and **b** Pt/Cu/HfO$_2$/TiN structures. E_F: Fermi level, CBM: conduction band minimum, VBM: valence band maximum

0.2 nm, respectively. The Pt-HfO$_2$ thickness is consistent with our previous report. The Cu$_2$O layer is less than 1 nm thick and should act as a metallic layer owing to the tunneling current and/or other leakage current via defects. In contrast, the Cu layer in TiN-HfO$_2$ was strongly oxidized. The Cu$_2$O layer may show an intrinsic semiconducting property since it is a p-type semiconductor [50]. The band alignments of Pt-HfO$_2$ and TiN-HfO$_2$ based on the above results and the reported band gaps of Cu$_2$O and HfO$_2$ [51, 52] are illustrated schematically in Fig. 4.17. For Pt-HfO$_2$, the Cu electrode acts as an electrode and the HfO$_2$ layer does not exhibit band bending. For TiN-HfO$_2$, there is a Cu$_2$O layer, which may act as a semiconducting layer, and a defective HfO$_2$ thin layer pinning the Fermi level, resulting in the band bending of the HfO$_2$ layer. The Cu/HfO$_2$ interface has the appearance of a p-n junction and exhibits nonlinear I–V properties.

4.6.3   Correlation Between Ion Migration and Switching Degradation

The I–V properties indicate the degradation of switching as the sweep number increases, as shown in Fig. 4.15. To investigate the effect of ion migration on degradation, HX-PES was performed. Pt-HfO$_2$ exhibited a reduction in the intensity of the Cu 2p$_{3/2}$ peak corresponding to Cu$_2$O, with increasing sweep number, indicating that the unintentionally oxidized layer was reduced continuously. For TiN-HfO$_2$, the

Fig. 4.18 Applied-voltage
sweep number dependences
of intensity ratios of Cu $2p_{3/2}$
to Hf $3d_{5/2}$ and O1s to total
intensity of Cu $2p_{3/2}$ and Hf
$3d_{5/2}$ normalized by Pt $4f_{7/2}$
intensity

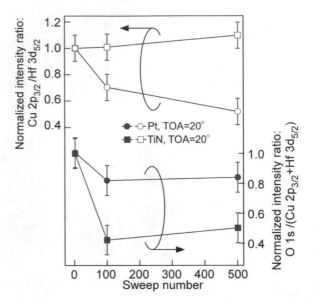

Fig. 4.18 Applied-voltage sweep number dependences of intensity ratios of Cu $2p_{3/2}$ to Hf $3d_{5/2}$ and O1s to total intensity of Cu $2p_{3/2}$ and Hf $3d_{5/2}$ normalized by Pt $4f_{7/2}$ intensity

relative intensity of the O 1s peak decreased. Furthermore, the Hf $3d_{5/2}$ peak showed a small change in its shape on the higher binding energy side, which was related to the defects at the interface. Figure 4.18 shows the intensity ratios of Cu $2p_{3/2}$ to Hf $3d_{5/2}$ (Cu/Hf) and O 1s to the total intensity of Cu $2p_{3/2}$ and Hf $3d_{5/2}$ (O/CuHf$_{total}$) plotted against the sweep number. For Pt-HfO$_2$, the Cu/Hf ratio decreased as the sweep number increased, suggesting that Cu diffused continuously into the HfO$_2$ layer. The O 1s peak intensity also decreased slightly. In contrast, for TiN-HfO$_2$, the Cu/Hf ratio was constant, and the O/CuHf$_{total}$ ratio decreased considerably, meaning that the oxygen moved to the opposite electrode side and that Cu migration did not occur.

4.6.4   Effect of Bottom Electrode on Conductive Filament Formation

At the Cu/HfO$_2$ interface, there is an unintentionally oxidized copper (Cu$_2$O) layer in both Pt-HfO$_2$ and TiN-HfO$_2$. For the Pt-HfO$_2$ structure, the application of a bias reduces the interfacial Cu$_2$O layer, which is a similar reaction to the forming process. The ionized Cu moves to the opposite electrode side with increasing sweep number, suggesting that the Cu density increased near the bottom electrode, which is consistent with the reported tendency. Ohno et al. reported that the endurance of the LRS of a nanoionics-type ReRAM structure depends on the sweep number and the turn on (set) voltage level, which is related to the density of Cu filaments [53, 54]. These results suggest that an effective way of controlling the endurance of nanoionics-type ReRAM is to control the total amount of Cu in the HfO_2 layer and the thickness of the Cu layer. For $TiN-HfO_2$, the Cu_2O layer is thicker than the Pt-HfO_2 layer at the Cu/HfO_2 interface. Furthermore, the HfO_2 layer of $TiN-HfO_2$ has many more oxygen vacancies than that of $Pt-HfO_2$. According to previous reports, a TiN electrode forms oxygen vacancies in the HfO_2 layer owing to the formation of an intermediate layer at the TiN/HfO_2 interface resulting from the oxidization of the TiN layer, and the asymmetrical structure enhances the oxygen vacancy distribution [55]. Furthermore, the incorporation of Hf enhances the oxidization of Cu [56]. The enhancement of oxygen migration by the formation of oxygen vacancies eliminated the reduction and ionization of Cu in the $Cu/HfO_2/TiN$ structure. These results suggest that the control of oxygen vacancy formation should help increase the endurance of Cu-electrode-based nanoionics-type ReRAM structures.

4.7   Summary

We have demonstrated the resistive switching of a $Cu/HfO_2/Pt$ structure whose materials and device structure are compatible with current LSI technology. We also found that HfO_2 thin films are polycrystalline with a columnar structure, making them suitable for Cu diffusion. The observed switching performance proved that HfO_2 is also a promising oxide for nonvolatile resistive switching applications, not simply as a high-k dielectric gate oxide. Furthermore, to investigate how to control the metal/oxide interface, we performed HX-PES under bias operation. For the Pt/HfO_2 interface, the application of a forward bias increased the Pt–O bonding peak intensity, indicating Pt electrode oxidization and oxygen vacancy formation around the interface. In contrast, the application of a bias to the Cu/HfO_2 interface reduced the copper oxide, providing evidence of oxygen reduction and Cu diffusion into the HfO_2 layer. The most plausible switching model of the Cu/HfO_2 interface was proposed on the basis of our experimental results. Cu ions diffuse in a solid electrolyte and undergo an electrochemical reaction at a cathode. By applying a positive voltage to an anode, Cu ions are generated around the anode and they diffuse toward the cathode. The fast diffusion of Cu into an oxide electrolyte means that it is easy to form filamentary paths in the electrolyte. On the other hand, ion–electron recombination occurs, causing metal segregation at the cathode. That is, the formation and annihilation conduction paths of metal atoms can be controlled by sweeping the voltage polarity.

HX-PES measurements also revealed the effect of a bottom electrode on a nanoionics-type $Cu/HfO_2/Pt$ ReRAM structure by comparison with the effect on a $Cu/HfO_2/TiN$ structure. In the case of a Pt bottom electrode, Cu reduction and migration were observed at the Cu/HfO_2 interface. As the switching number increased, the Cu moved toward the bottom electrode. In contrast, a TiN bottom electrode induced oxygen vacancy formation in the HfO_2 layer and the oxidization of the Cu layer, resulting in the Cu/HfO_2 interface having an accumulation layer, and oxygen migrated in the HfO_2 layer instead of the Cu layer. These results suggest that the density of oxygen vacancies is important in terms of controlling the formation of conductive filaments in a nanoionics-type ReRAM structure.

References

1. Waser R (2009) Resistive non-volatile memory devices. Microlelectron Eng 86:1925–1928. https://doi.org/10.1016/j.mee.2009.03.132
2. Karg SF, Meijer GI, Bednorz JG, Rettner CT, Schtorr AG, Joseph EA, Lam CH, Janousch M, Staub U, La Mattina F, Alvarado SF, Widmer D, Stutz R, Drechsler U, Caimi D (2008) Transition-metal-oxide-based resistance-change memories. IBM J Res Dev 52:481–492. https://doi.org/10.1147/rd.524.0481
3. Pagnia H, Sotnik N (1988) Bistable switching in electroformed metal–insulator–metal devices. Phys Status Solidi 108:11–65. https://doi.org/10.1002/pssa.2211080102
4. Chudnovskii FA, Odynets LL, Pergament AL, Stefanovich GB (1996) Electroforming and switching in oxides of transition metals: the role of metal–insulator transition in the switching mechanism. J Solid State Chem 122:95–99. https://doi.org/10.1006/jssc.1996.0087
5. Asamitsu A, Tomioka Y, Kuwahara H, Tokura Y (1997) Current switching of resistive states in magnetoresistive manganites. Nature 388:50. https://doi.org/10.1038/40363
6. Fors R, Khartsev SI, Grishin AM (2005) Giant resistance switching in metal-insulator-manganite junctions: evidence for Mott transition. Phys Rev B 71:045305. https://doi.org/10.1103/PhysRevB.71.045305
7. Kim DS, Kim YH, Lee CE, Kim YT (2006) Colossal electroresistance mechanism in a $Au/Pr_{0.7}Ca_{0.3}MnO_3/Pt$ sandwich structure: evidence for a Mott transition. Phys Rev B 74:174430. https://doi.org/10.1103/physrevb.74.174430
8. Meijer GI, Staub U, Janousch M, Johnson SL, Delley B, Neisius T (2005) Valence states of Cr and the insulator-to-metal transition in Cr-doped $SrTiO_3$. Phys Rev B 72:155102. https://doi.org/10.1103/PhysRevB.72.155102
9. Waser R, Aono M (2007) Nanoionics-based resistive switching memories. Nat Mater 6:833–840. https://doi.org/10.1038/nmat2023
10. Sakamoto T, Sunamura H, Kawaura H, Hasegawa T, Nakayama T, Aono M (2003) Nanometer-scale switches using copper sulfide. Appl Phys Lett 82:3032. https://doi.org/10.1063/1.1572964
11. Banno N, Sakamoto T, Hasegawa T, Terabe K, Aono M (2006) Effect of ion diffusion on switching voltage of solid-electrolyte nanometer switch. Jpn J Appl Phys 46:3666–3668. https://doi.org/10.1143/JJAP.45.3666
12. Terabe K, Hasegawa T, Nakayama T, Aono M (2005) Quantized conductance atomic switch. Nature 433:47–50. https://doi.org/10.1038/nature03190
13. Kozicki MN, Park M, Mitkova M (2005) Nanoscale memory elements based on solid-state electrolytes. IEEE Trans Nanotechnol 4:331. https://doi.org/10.1109/TNANO.2005.846936
14. Kim YM, Lee JS (2008) Reproducible resistance switching characteristics of hafnium oxide-based nonvolatile memory devices. J Appl Phys 104:114115. https://doi.org/10.1063/1.3041475
15. Sakamoto T, Lister K, Banno N, Hasegawa T, Terabe K, Aono M (2007) Electronic transport in Ta_2O_5 resistive switch. Appl Phys Lett 91:92110. https://doi.org/10.1063/1.2777170
16. Lee S, Kim WG, Rhee SW, Yong K (2007) Resistance switching behaviors of hafnium oxide films grown by MOCVD for nonvolatile memory applications. J Electrochem Soc 155:H92–H96. https://doi.org/10.1149/1.2814153
17. Gibbons JF, Beadle WE (1964) Switching properties of thin NiO films. Solid-State Electron 7:785–790. https://doi.org/10.1016/0038-1101(64)90131-5

18. Tsuchiya T, Oyama Y, Miyoshi S, Yamaguchi S (2009) Nonstoichiometry-induced carrier modification in gapless type atomic switch device using Cu_2S mixed conductor. Appl Phys Express 2:055002. https://doi.org/10.1143/APEX.2.055002
19. Haemori M, Nagata T, Chikyow T (2009) Impact of Cu electrode on switching behavior in a $Cu/HfO_2/Pt$ structure and resultant Cu ion diffusion. Appl Phys Express 2:061401. https://doi.org/10.1143/APEX.2.061401
20. Wilk GD, Wallace RM (1999) Electrical properties of hafnium silicate gate dielectrics deposited directly on silicon. Appl Phys Lett 74:2854. https://doi.org/10.1063/1.124036
21. Robertson J (2006) High dielectric constant gate oxides for metal oxide Si transistors. Rep Prog Phys 69:327. https://doi.org/10.1088/0034-4885/69/2/R02
22. Yang JJ, Miao F, Pickett MD, Ohlberg DAA, Stewart DR, Lau CN, Williams RS (2009) The mechanism of electroforming of metal oxide memristive switches. Nanotechnology 20:215201. https://doi.org/10.1088/0957-4484/20/21/215201
23. Yoshida C, Kinoshita K, Yamasaki T, Sugiyama Y (2008) Direct observation of oxygen movement during resistance switching in NiO/Pt film. Appl Phys Lett 93:042106. https://doi.org/10.1063/1.2966141
24. Shima H, Takano F, Muramatsu H, Yamazaki M, Akinaga H, Kogure A (2008) Local chemical state change in Co–O resistance random access memory Phys Status Solidi 2:99–101. https://doi.org/10.1002/pssr.200802003
25. Yamashita Y, Ohmori K, Ueda S, Yoshikawa H, Chikyow T, Kobayashi K (2010) Bias-voltage application in hard x-ray photoelectron spectroscopy for characterization of advanced materials. e-J Surf Sci Nanotechnol 8:81–83. https://doi.org/10.1380/ejssnt.2010.81
26. Nagata T, Haemori M, Yamashita Y, Iwashita Y, Yoshikawa H, Kobayashi K, Chikyow T (2010) Oxygen migration at $Pt/HfO_2/Pt$ interface under bias operation. Appl Phys Lett 97:082902. https://doi.org/10.1063/1.3483756
27. Tsuruoka T, Terabe K, Hasegawa T, Aono M (2010) Forming and switching mechanisms of a cation-migration-based oxide resistive memory. Nanotechnology 21:425205. https://doi.org/10.1088/0957-4484/21/42/425205
28. Yang JJ, Pickett MD, Li X, Ohlberg DAA, Stewart DR, Williams RS (2008) Memristive switching mechanism for metal/oxide/metal nanodevices. Nat Nanotechnol 3:429–433. https://doi.org/10.1038/nnano.2008.160
29. Schroeder H, Jeong DS (2007) Resistive switching in a $Pt/TiO_2/Pt$ thin film stack—a candidate for a non-volatile ReRAM. Microelectron Eng 84:1982–1985. https://doi.org/10.1016/j.mee.2007.04.042
30. Poulston S, Parlett PM, Stone P, Bowker M (1996) Surface oxidation and reduction of CuO and Cu_2O studied using XPS and XAES. Surf Interface Anal 24:811–820. https://doi.org/10.1002/(SICI)1096-9918(199611)24:12%3c811:AID-SIA191%3e3.0.CO;2-Z
31. Galtayrise A, Bonnelle JP (1995) XPS and ISS studies on the interaction of H_2S with polycrystalline Cu, Cu_2O and CuO surfaces. Surf Interface Anal 23:171–179. https://doi.org/10.1002/sia.740230308
32. Rhodin TN Jr (1950) Low temperature oxidation of copper. I. Physical mechanism. J Am Chem Soc 72:5102–5106. https://doi.org/10.1021/ja01167a079
33. Iijima J, Lim JW, Hong SH, Suzuki S, Mimura K, Isshiki M (2006) Native oxidation of ultra high purity Cu bulk and thin films. Appl Surf Sci 253:2825–2829. https://doi.org/10.1016/j.apsusc.2006.05.063
34. Himpsel FJ, McFeely FR, Taleb-Ibrahimi A, Yarmoff JA, Hollinger G (1988) Microscopic structure of the SiO_2/Si interface. Phys Rev B 38:6084–6096. https://doi.org/10.1103/PhysRevB.38.6084
35. Kobayashi K, Yabashi M, Takata Y, Tokushima T, Shin S, Tamasaku K, Miwa D, Ishikawa T, Nohira H, Hattori T, Sugita Y, Nakatsuka O, Sakai A, Zaima S (2003) High resolution-high energy x-ray photoelectron spectroscopy using third-generation synchrotron radiation source, and its application to Si-high k insulator systems. Appl Phys Lett 83:1005. https://doi.org/10.1063/1.1595714

36. Barreca D, Milanov A, Fischer RA, Devi A, Tondello E (2007) Hafnium oxide thin film grown by ALD: an XPS study. Surf Scie Spectra 14:34. https://doi.org/10.1116/11.20080401
37. Paàl Z, Muhler M, Schlögl R (1996) Platinum black by XPS. Surf Scie Spectra 4:119. https://doi.org/10.1116/1.1247817
38. Jung M-C, Kim H-D, Han M, Jo W, Kim DC (1999) X-ray photoelectron spectroscopy study of Pt-oxide thin films deposited by reactive sputtering using O_2/Ar gas mixtures. Jpn J Appl Phys 38:4872. https://doi.org/10.1143/JJAP.38.4872
39. Matolín V, Cabala M, Matolínová I, Škoda M, Václavů M, Prince KC, Skála T, Mori T, Yoshikawa H, Yamashita Y, Ueda S, Kobayashi K (2010) Pt and Sn doped sputtered CeO_2 electrodes for fuel cell application. Fuel Cells 10:139–144. https://doi.org/10.1002/fuce.200900036
40. Bard AJ, Faulkner LR (2001) Electrochemical methods. Fundamentals and applications, 2nd edn. Wiley, New York
41. Yoshitake M, Aparna YR, Yoshihara K (2001) General rule for predicting surface segregation of substrate metal on film surface. J Vac Sci Technol A 19:1432. https://doi.org/10.1116/1.1376699
42. Takeuchi H, Ha D, King TJ (2004) Observation of bulk HfO_2 defects by spectroscopic ellipsometry. J Vac Sci Technol A 22:1337. https://doi.org/10.1116/1.1705593
43. Ohmori K, Ahmet P, Yoshitake M, Chikyow T, Shiraishi K, Yamabe K, Watanabe H, Akasakad Y, Nara Y, Chang KS, Green ML, Yamada K (2007) Influences of annealing in reducing and oxidizing ambients on flatband voltage properties of HfO_2 gate stack structures. J Appl Phys 101:084118. https://doi.org/10.1063/1.2721384
44. Lin K-L, Hou T-H, Shieh J, Lin J-H, Chou C-T, Lee Y-J (2011) Electrode dependence of filament formation in HfO_2 resistive-switching memory. J Appl Phys 109:084104. https://doi.org/10.1063/1.3567915
45. Cagli C, Buckley J, Jousseaume V, Cabout T, Salaun A, Grampeix H, Nodin JF, Feldis H, Persico A, Cluzel J, Lorenzi P, Massari L, Rao R, Irrera F, Aussenac F, Carabasse C, Coue M, Calka P, Martinez E, Perniola L, Blaise P, Fang Z, Yu YH, Ghibaudo G, Deleruyelle D, Bocquet M, Müller C, Padovani A, Pirrotta O, Vandelli L, Larcher L, Reimbold G, de Salvo B (2011) Experimental and theoretical study of electrode effects in HfO_2 based RRAM. In: IEDM 2011 proceedings, p 658. https://doi.org/10.1109/IEDM.2011.6131634
46. Kamiya K, Yang MY, Nagata T, Park S-G, Magyari-Kope B, Chikyow T, Yamada K, Niwa M, Nishi Y, Shiraishi K (2013) Generalized mechanism of the resistance switching in binary-oxide-based resistive–random–access–memories. Phys Rev B 87:155201. https://doi.org/10.1103/PhysRevB.87.155201
47. Goux L, Czarnecki P, Chen YY, Pantisano L, Wang XP, Degraeve R, Govoreanu B, Jurczak M, Wouters DJ, Altimime L (2010) Evidences of oxygen-mediated resistive-switching mechanism in TiN/HfO_2/Pt cells. Appl Phys Lett 97:243509. https://doi.org/10.1063/1.3527086
48. Nagata T, Oh S, Yamashita Y, Yoshikawa H, Hayakawa R, Kobayashi K, Chikyow T, Wakayama Y (2012) Hard x-ray photoelectron spectroscopy study on band alignment at poly(3,4-ethylenedioxythiophene):poly(styrenesulfonate)/ZnO interface. Appl Phys Lett 101:173303. https://doi.org/10.1063/1.4762834
49. http://www.sasj.jp/COMPRO
50. Raebiger H, Lany S, Zunger A (2007) Origins of the p-type nature and cation deficiency in Cu_2O and related materials. Phys Rev B 76:045209. https://doi.org/10.1103/PhysRevB.76.045209
51. Tahir D, Tougaard S (2012) Electronic and optical properties of Cu, CuO and Cu_2O studied by electron spectroscopy. J Phys Condens Mater 24:175002. https://doi.org/10.1088/0953-8984/24/17/175002
52. Robertson J (2000) Band offsets of wide-band-gap oxides and implications for future electronic devices. J Vac Sci Technol, B 18:1785. https://doi.org/10.1116/1.591472
53. Ohno T, Hasegawa T, Tsuruoka T, Terabe K, Gimzewski JK, Aono M (2011) Short-term plasticity and long-term potentiation mimicked in single inorganic synapses. Nat Mater 10:591. https://doi.org/10.1038/NMAT3054

54. Ohno T, Hasegawa T, Nayak A, Tsuruoka T, Gimzewski JK, Aono M (2011) Sensory and short-term memory formations observed in a Ag_2S gap-type atomic switch. Appl Phys Lett 99:203108. https://doi.org/10.1063/1.3662390

55. Sowinska M, Bertaud T, Walczyk D, Thiess S, Schubert MA, Lukosius M, Drube W, Walczyk C, Schroeder T (2012) Hard x-ray photoelectron spectroscopy study of the electroforming in Ti/HfO_2-based resistive switching structures. Appl Phys Lett 100:233509. https://doi.org/10.1063/1.4728118

56. Tam CY, Shek CH (2005) Effects of alloying on oxidation of Cu-Based bulk metallic glasses. J Mater Res 20:2647. https://doi.org/10.1557/JMR.2005.0336

Chapter 5
Switching Control of Oxide-Based Resistive Random-Access Memory by Valence State Control of Oxide

5.1 Introduction

In the previous chapter, we described the investigation of suitable electrode and matrix materials for the LSI process and demonstrated the resistance switching behavior of the Cu/HfO$_2$/Pt system [1–3]. There has been increased interest in the use of high-k dielectric oxides as potential ReRAM materials. Moreover, switching voltage control is beneficial for actual applications. In terms of controlling the operating voltage, the matrix materials are critical for controlling the diffusion of Cu. Metal ions diffuse in a solid electrolyte through defects, including oxygen vacancies or grain boundaries, and an electrochemical reaction at another electrode, meaning that the matrix materials affect the switching voltage.

In this chapter, we discuss the Ta–Nb binary oxide (Ta$_x$Nb$_{1-x}$)$_2$O$_5$ system using a combinatorial synthesis technique for controlling the metal ion diffusion properties induced by oxygen vacancies and the valence of the metals.

5.2 Valence Control Scheme

Ta$_2$O$_5$ and Nb$_2$O$_5$ are candidate high-k materials and have high affinity in the LSI process. The physical properties of Ta and Nb are similar in terms of ion radius, electronegativity, and oxidation number, although they differ in terms of Gibbs free energy for oxidization (ΔG). On the basis of the Ellingham diagram shown in Fig. 5.1 [4], tantalum oxide (TaO$_x$) typically has two different valence states with the following ΔG values at room temperature: Ta$_2$O$_5$ (Ta^{5+}) $\Delta G = -760.5$ kJ mol^{-1} and TaO (Ta^{2+}) $\Delta G = -54.125$ kJ mol^{-1}. In contrast, niobium oxide (NbO$_x$) typically has three different valence states with the following ΔG values at room temperature: Nb$_2$O$_5$ (Nb^{5+}), $\Delta G = -704.8$ kJ mol^{-1}; NbO$_2$ (Nb^{4+}), $\Delta G = -733.5$ kJ mol^{-1}; and NbO (Nb^{2+}), $\Delta G = -773.3$ kJ mol^{-1}. These energy differences can be expected to induce a gradient in the ion diffusion properties of the (Ta$_x$Nb$_{1-x}$)$_2$O$_5$ system.

© National Institute for Materials Science, Japan 2020
T. Nagata, *Nanoscale Redox Reaction at Metal/Oxide Interface*, NIMS Monographs,
https://doi.org/10.1007/978-4-431-54850-8_5

Fig. 5.1 Ellingham diagram
for Ta, Cu, Nb, and Hf oxides

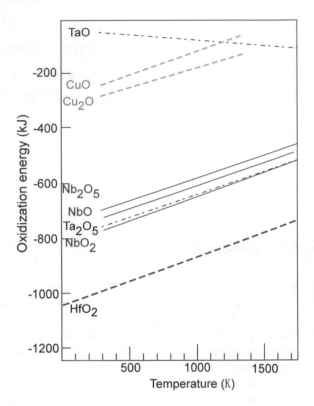

5.3 Combinatorial Synthesis

5.3.1 Ta–Nb Binary Oxide System

A 100-nm-thick Pt bottom electrode layer was deposited on a Si substrate by DC
sputtering at room temperature. A Ta–Nb binary oxide film was deposited on the
Pt bottom electrode by combinatorial pulsed laser deposition (combi-PLD) using
procedures similar to binary film deposition, as shown in Chap. 6 (Fig. 6.2).

Figure 5.2a, b shows a schematic illustration of a sample and the results of metal
composition measurements by XRF, respectively. The XRF measurements revealed
that the Ta and Nb contents change continuously, suggesting that a composition
spread film sample was obtained. The 2D-XRD measurements revealed that the
Ta_2O_5 region exhibited no diffraction pattern, indicating an amorphous structure.
For the Nb_2O_5 region, the 2D-XRD image showed additional weak ring patterns,
which were confirmed as the Nb_2O_5 structure. These results revealed that the region
included a small polycrystalline phase. The other Ta–Nb binary regions showed the
same 2D-XRD patterns as the Ta_2O_5 region, indicating amorphous structures.

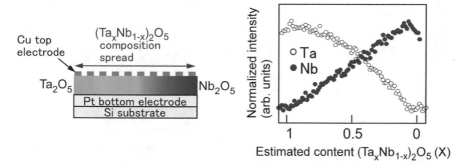

Fig. 5.2 a Schematic illustration and **b** composition map measured by XRF of Cu/(Ta$_x$Nb$_{1-x}$)$_2$O$_5$/Pt structure

5.3.2   Valence State of Oxides

XPS measurements were performed to investigate the valence state of the metals and the chemical bonding state of oxygen. Figure 5.3 shows XPS intensity map for Nb 3d and Ta 4f core-level spectra. In the Ta spectra for Ta$_2$O$_5$, Ta 4f clearly shifted by approximately 0.50 eV toward a lower binding energy when a small amount of Nb was added. The binding energy for the TaO bonding state was approximately 2.4 eV lower than that for Ta$_2$O$_5$ [5]. Furthermore, a tail state denoted by an asterisk in Fig. 5.4a was observed. In Fig. 5.4a, to compare the shape of the spectra, the spectrum of (Ta$_{0.8}$Nb$_{0.2}$)$_2$O$_5$ was offset by +0.5 eV along the x-axis. The inset shows a difference peak that is the peak of Ta$_2$O$_5$ minus that of (Ta$_{0.8}$Nb$_{0.2}$)$_2$O$_5$, which is attributed to the defect state in Ta$_2$O$_5$ [6]. These results suggested a valence number of 5+ (Ta$_2$O$_5$) for Ta and the existence of a defect tail state, which is related to both the oxygen vacancy and the Ta defect. In contrast, with Nb 3d, the changing

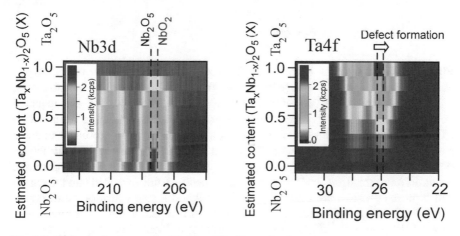

Fig. 5.3 XPS mapping images for Nb 3d and Ta 4f core level spectra

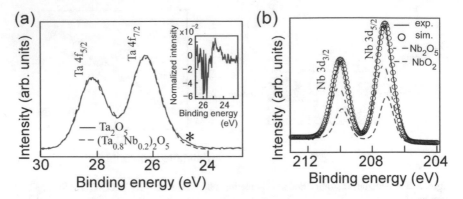

Fig. 5.4 **a** XPS for Ta 4f of Ta_2O_5 (solid line), and $(Ta_{0.8}Nb_{0.2})_2O_5$ (dashed line). To compare the spectrum shape, the spectrum of $(Ta_{0.8}Nb_{0.2})_2O_5$ was shifted by 0.5 eV. The inset shows the difference peak, which is the Ta 4f peak of Ta_2O_5 minus that of $(Ta_{0.8}Nb_{0.2})_2O_5$. **b** XPS for Nb 3d of $(Ta_{0.2}Nb_{0.5})_2O_5$. The solid lines and open circles show the experimental spectrum and sum-fitted curve, respectively. The dashed lines are fitted curves for each component: Nb_2O_5 and NbO_2

behavior of the spectral shape was different from that of Ta 4f. The asymmetrical shape became pronounced as the Ta content increased and could be deconvoluted, as shown in Fig. 5.4b. The Nb $3d_{5/2}$ spectra at 207.29 and 207.09 eV can be assigned to the Nb_2O_5 and NbO_2 components, respectively, indicating that the valence state of Nb was changed by adding Ta [7, 8]. In the O 1 s spectra, the total amount of oxygen increased with increasing Ta content. It can be assumed that the Nb_2O_5 region has a higher oxygen vacancy density than Ta_2O_5. Additionally, the valence of Nb 4+ requires less oxygen than the valence of Nb 5+, meaning that the $(Ta_xNb_{1-x})_2O_5$ region should have fewer oxygen vacancies than Nb_2O_5. The oxygen vacancies can be compensated for by controlling the valence of Nb.

5.3.3 Electrical Properties

Figure 5.5 shows typical I–V characteristics for several compositions. The resistance changed when an applied voltage was swept from a positive bias to a negative bias, indicating that there are two different resistive states in the positive and negative voltage regions; one is a high-resistance state (HRS: turn-off) and the other is a low-resistance state (LRS: turn-on). In the Ta_2O_5 and Nb_2O_5 regions, the I–V properties showed poor leakage properties at the initial state, which corresponded to the Ta defect and/or oxygen vacancy state as observed in the XPS results. Although the saturation current decreased markedly as a result of the first formation (turn-on and turn-off) process, the turn-off process exhibited a step structure, as shown in Fig. 5.5a, f. One possible reason is the Joule heating effect on the oxygen vacancies in the LRS. Currently, this step structure cannot be controlled, meaning that the oxygen vacancy potentially affects the switching process and makes the formation

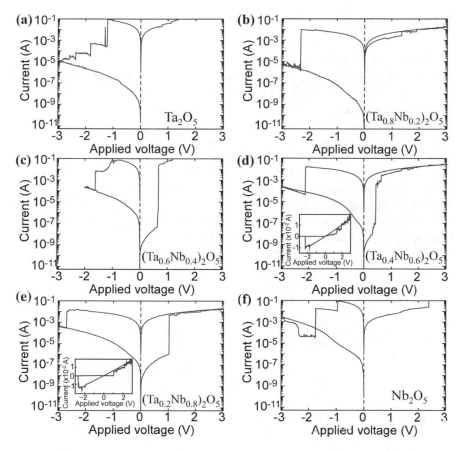

Fig. 5.5 I–V characteristics of **a** Ta_2O_5, **b** $(Ta_{0.8}Nb_{0.2})_2O_5$, **c** $(Ta_{0.6}Nb_{0.4})_2O_5$, **d** $(Ta_{0.4}Nb_{0.6})_2O_5$, **e** $(Ta_{0.2}Nb_{0.6})_2O_5$, and **f** Nb_2O_5. The current compliance was set at 100 mA

process unstable. Within the composition spread region, and especially the Nb-rich region (Fig. 5.5d, e), the initial leakage properties are better than those of the Ta_2O_5 and Nb_2O_5 regions. The LRS is lower and more stable than in the other regions. Furthermore, in the turn-off process, the I–V property showed a single-step structure. The insets in Fig. 5.5d, e show plots of the I–V characteristic, which indicated linearity. Note that the oxygen vacancy model relating to the oxygen migration at the metal/oxide interface shows nonlinear I–V curves during its formation [9, 10], meaning that, with $Cu/(Ta_xNb_{1-x})_2O_5$, the majority of the conducting paths at the interface should be metal-conducting paths in the oxide (nanoionics model). By combining the XPS and I–V results, it can be suggested that the $(Ta_xNb_{1-x})_2O_5$ having high Nb-content ($0 < x < 0.5$) regions with the Cu top electrode is a good candidate for ReRAM applications.

5.4 Summary

In summary, we have used combinatorial synthesis to control the resistive switching properties of the $Cu/(Ta_xNb_{1-x})_2O_5/Pt$ structure, whose materials and device structure are compatible with current LSI technology. The as-deposited $(Ta_xNb_{1-x})_2O_5$ film with a grain structure showed resistive switching behavior. XPS revealed that the valence of Nb and the oxygen vacancies were affected by the Ta content. As regards the I–V properties, the valence in the 5+ region did not exhibit a stable resistive changing behavior in contrast to the low-Ta-content region. In particular, $(Ta_xNb_{1-x})_2O_5$ with a Nb content above 0.5 did not require thermal treatment to obtain initial low leakage current properties, which can reduce the damage caused by the Joule heating effect. These results suggest that doping Nb_2O_5 with Ta makes it possible to control the valence of Nb and the resistive changing behavior of the $Cu/(Ta_xNb_{1-x})_2O_5$ nanoionic-type ReRAM structure.

References

1. Haemori M, Nagata T, Chikyow T (2009) Impact of Cu electrode on switching behavior in a Cu/HfO$_2$/Pt structure and resultant Cu ion diffusion. Appl Phys Express 2:061401. https://doi.org/10.1143/APEX.2.061401
2. Nagata T, Haemori M, Yamashita Y, Iwashita Y, Yoshikawa H, Kobayashi K, Chikyow T (2010) Oxygen migration at Pt/HfO$_2$/Pt interface under bias operation. Appl Phys Lett 97:082902. https://doi.org/10.1063/1.3483756
3. Nagata T, Haemori M, Yamashita Y, Yoshikawa H, Iwashita Y, Kobayashi K, Chikyow T (2011) Bias application hard X-ray photoelectron spectroscopy study of forming process of Cu/HfO$_2$/Pt resistive random access memory structure. Appl Phys Lett 99:223517. https://doi.org/10.1063/1.3664781
4. Ellingham HJT (1944) Transactions and communications. J Soc Chem Ind (London) 63:125–160. https://doi.org/10.1002/jctb.5000630501
5. Kerrec O, Devilliers D, Groult H, Marcus P (1998) Study of dry and electrogenerated Ta2O5 and Ta:Ta2O5: Pt structures by XPS. Mat Sci Eng B55:134–142. https://doi.org/10.1016/S0921-5107(98)00177-9
6. Kohli S, McCurdy PR, Rithner CD, Dorhout PK, Dummer AM, Brizuela F, Menoni CS (2004) X-ray characterization of oriented β-tantalum films. Thin Solid Films 469:404–409. https://doi.org/10.1016/j.tsf.2004.09.001
7. Fontaine R, Caillat R, Feve L, Guittet MJ (1977) Déplacement chimique ESCA dans la série des oxydes du niobium. J Electron Spectrosc Relat Phenom 10:349–357. https://doi.org/10.1016/0368-2048(77)85032-9
8. Simon D, Perrin C, Baillif P (1976) Electron spectrometric study (ESCA) of niobium and its oxides. Application to its oxidation at high temperature and low oxygen pressure. C R Acad Sci Ser C 283:241–244
9. Yang JJ, Pickett MD, Li X, Ohlberg DAA, Stewart DR, Williams RS (2008) Memristive switching mechanism for metal/oxide/metal nanodevices. Nat Nanotechnol 3:429–433. https://doi.org/10.1038/nnano.2008.160
10. Schroeder H, Jeong DS (2007) Resistive switching in a Pt/TiO$_2$/Pt thin film stack—a candidate for a nonvolatile ReRAM. Microelec Eng 84:1982–1985. https://doi.org/10.1016/j.mee.2007.04.042

Chapter 6
Combinatorial Thin-Film Synthesis for New Nanoelectronics Materials

6.1 Introduction

The combinatorial synthesis technique is increasingly commonly used for optimizing numerous multicomponent functional materials [1–4]. Although the combinatorial synthesis technique based on pulsed laser deposition (PLD) is superior for screening oxide materials, it is frequently unsuccessful when applied to metal thin films owing to droplet formation during deposition [5]. More recently, multicathode sputtering systems have been used for composition spread metal film fabrication [6]. This system, however, has the drawbacks of controllability of composition spread owing to its broad sputtering erosion and high-energy plasma. To overcome this problem, we have proposed combinatorial synthesis using focused Ar ion-beam sputtering (FIBS) [7]. Ion-beam sputtering is suitable for metal deposition, since it does not result in droplet formation, and can be applied to almost every kind of material. Although a conventional ion-beam sputtering system uses a high-energy ion beam, our system uses a low-energy focused Ar ion beam source with a spot size of 1 mm diameter on a target. Using this ion source, we can use the same alignment and multitarget system as a PLD system.

6.2 Combinatorial Thin-Film Synthesis

Figure 6.1 shows a typical setup of a combinatorial PLD system. For oxides, a UV laser such as a KrF excimer laser ($\lambda = 248$ nm) or a Q switched Nd:YAG laser is used. In the case of FIBS, a focused Ar ion beam gun is used instead of the laser. The system has a moving mask, a target exchange system, and a substrate rotation system, all of which are controlled using software on a PC [8]. For the combinatorial thin-film synthesis, a uniform composition spread film is required. The composition and thickness of the film can be precisely controlled using the moving mask. To adjust the amount of target material and give the same thickness in the film, the mask

© National Institute for Materials Science, Japan 2020
T. Nagata, *Nanoscale Redox Reaction at Metal/Oxide Interface*, NIMS Monographs,
https://doi.org/10.1007/978-4-431-54850-8_6

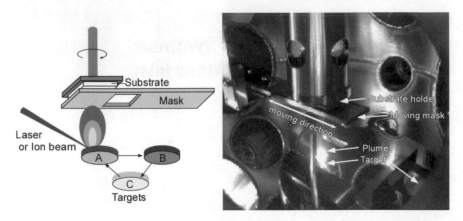

Fig. 6.1 Schematic illustration and photograph of combinatorial thin-film deposition system

speed is calibrated. For a fast-growth material, the mask moves faster than that for a slow-growth material. Figure 6.2 shows a binary composition spread deposition cycle, which consists of two steps (i and ii). In each step, a mask with a square-shaped hole moves at a constant speed from one side of the substrate to the other, after which

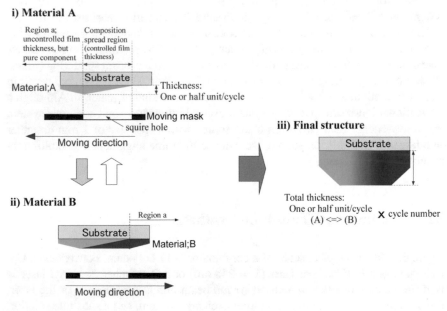

Fig. 6.2 Schematic illustration of composition-spread thin-film sample fabrication procedures, a binary composition-spread deposition cycle. There were two steps: (i) material A deposition and (ii) material B deposition. In each deposition, a mask moved at constant speed from one side of the substrate to the other, after which the targets were changed. Alternating between steps (i) and (ii) created binary composition-spread samples (iii)

the targets are changed. The maximum film thickness in one cycle is typically one or one-half unit cell of materials. Alternating between steps (i) and (ii) created the composition spread region. In the case of a ternary system, two different methods are used. One is the substrate rotation method, which was developed by Chikyow and colleagues [9, 10], as shown in Fig. 6.3. A substrate is rotated by 120° after the deposition of each material. The advantage of this method is the accuracy of the composition of each component. The other is the optimized mask method, which was developed by Matsumoto and colleagues [11, 12], as shown in Fig. 6.4. Using an optimized mask for ternary composition spread sample fabrication, we can fabricate the sample more easily than by the substrate rotation method.

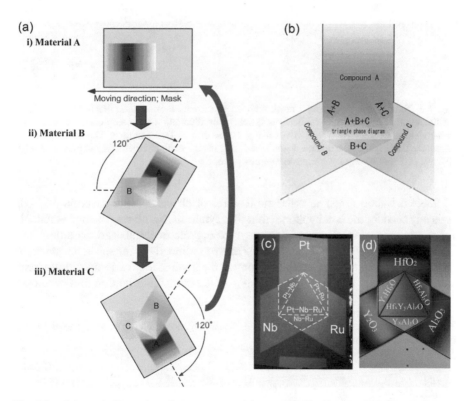

Fig. 6.3 **a** Schematic illustration of ternary composition-spread thin-film sample fabrication procedures. There were three steps: (i) material A deposition, (ii) material B deposition, and (iii) material C deposition. In each deposition, after rotating a substrate by 120°, a mask moved at constant speed from one side of the substrate to the other, after which the targets were changed. Alternating between steps (i), (ii), and (iii) created ternary composition-spread samples. **b** Schematic illustration of ternary composition-spread thin-film sample structure. **c** Photograph of a ternary alloy thin-film sample. **d** Photograph of a ternary oxide composition-spread thin-film sample

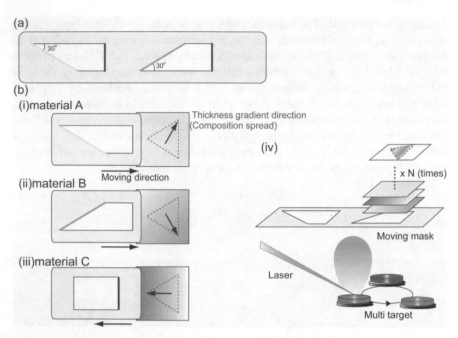

Fig. 6.4 Ternary alloying using a mask with linear motion. **a** Schematic illustration of a mask for the ternary alloy thin-film fabrication. **b** Schematic illustration of ternary composition-spread thin-film sample fabrication procedures using the mask with linear motion. The edge angle and length are the origin of the thickness gradient during the deposition. Alternating between steps (i), (ii), and (iii) created ternary composition-spread samples (iv)

Table 6.1 shows the characteristic features of physical vapor devotion methods typically used for the combinatorial thin-film synthesis. Sputtering method is suitable for the deposition of all materials except for organic materials and mass products. However, some of the sputtered particles move nonlinearly to the substrate and come around behind the moving mask, resulting in the nonlinearity of composition spread sample formation. From the point of view of the combinatorial thin film synthesis,

Table 6.1 Characteristic features of physical vapor devotion methods typically used for the combinatorial thin film synthesis

	Metal deposition	Oxide/Nitride deposition	Organic film deposition	large-area deposition	Fast growth rate	Composition Linearity	Handleability /Maintenance	Kinetic energy	Co-deposition
MBE	S	A	S	C	C	S	C	S	S
PLD	C	S	B	C	B	S	A	B	C
FIBS	S	A	C	C	C	S	A	A	C
Sputtering	S	S	C	S	S	B	S	C	A

S: Very Good, A: Good, B:Poor, C:Bad

a slow deposition rate and the linear move of deposited particle are important to obtain the linear composition spread sample. MBE, PLD, and FIBS are suitable for the combinatorial thin-film synthesis although each method has strong materials and weak materials.

6.3 Focused Ar Ion-Beam Sputtering for Combinatorial Synthesis

6.3.1 Energy of Focused Ar Ion Beam Sputtering

To obtain a fine-structured combinatorial thin-film sample, the controllability of the deposition rate and the deposited particle energy are important. In the case of the FIBS, the energy of a deposited particle was determined by Langmuir probe measurements [13].

The base pressure was 7×10^{-8} Pa, and the Ar partial pressure during deposition was 7×10^{-6} Pa. The ion current with 5 kV acceleration energy at a target was 40 μA. Langmuir probe measurements were conducted using a tungsten wire of 0.25 mm diameter protruding 3 mm from a ceramic sleeve and mounted just beneath the substrate holder, as shown in Fig. 6.5a. The distance between the substrate holder and the probe was 0.5 mm. The bias voltage (V) applied to the probe was in the range of −60 to 60 V, with the ionic current from the plasma measured using a picoammeter. Figure 6.5b shows Langmuir probe current–voltage (I–V) plots for a Pt deposition by FIBS. For negative probe biases, electrons are repelled from the probe. The floating potential (V_f) is defined as the point where the ionic current is zero. Around V_f, the slope of the curve begins to increase, and both ions and electrons can reach the probe. The plasma potential V_s is the point above which the ion flux decreases until

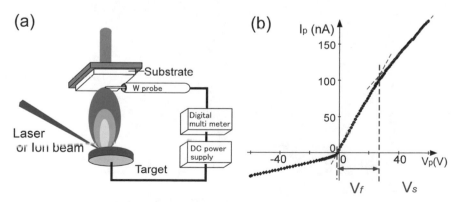

Fig. 6.5 **a** Experimental setup for Langmuir probe measurement. **b** Langmuir probe I–V curves for Pt target

only electrons can reach the probe. It is also marked by a change in the slope of the current–voltage curve.

The ion density (N_i) can be deduced from the slope of the I^2–V curve in the region between the floating and plasma potentials. In this study, we assumed that the plasma is collisionless, the electron energy distribution is Maxwellian, and the electron temperature is much higher than the ion temperature. It was assumed that N_i is equal to the electron density (N_e) because the plasma is neutral. N_i is obtained using the following equation [14, 15].

$$N_i^2 = N_e^2 = \frac{4\pi m_e}{3A^2 e^3}\left(\frac{\partial I_e^2}{\partial V_p}\right), \tag{6.1}$$

where m_e is the electron mass, A is the probe area, and the probe voltage V_p is in the range where random ion and electron currents are unaffected by bias voltage (i.e., between the floating and plasma potentials). The electron temperature (T_e) is determined from the slope of the $\ln(I)$–V_p curve of the probe using the equation

$$T_e = \frac{\partial V_p}{\partial \ln(I_e)}, \tag{6.2}$$

where I_e is the electron current.

The values of N_i and T_e of the Pt deposition were 9.8×10^{12} m^{-3} and 2.8 eV, respectively. Figure 6.6 shows an energy diagram of N_i and T_e obtained by various deposition methods. In conventional sputtering, the range of 10^{15}–10^{18} m^{-3} was

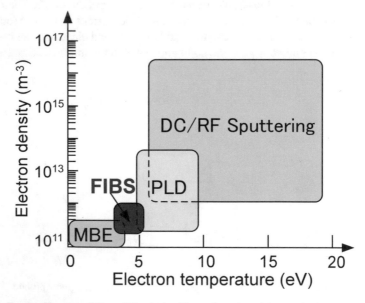

Fig. 6.6 Energy diagram of N_i and T_e obtained by various deposition methods

reported at a point 5–10 cm above the target in Ar sputtering plasma [16, 17]. The values of N_i and T_e measured in the plumes of GaN PLD were 3.72×10^{14} m^{-3} and 5.54 eV, respectively [18]. The energy of FIBS is lower than that of these methods and reaches the energy of the molecular beam epitaxy (MBE) method. In addition, FIBS does not require any liquid nitrogen supply or an especially complex control system such as an MBE system.

6.3.2 Metal Thin-Film Growth on Oxide

FIBS can be used to deposit fine and low-damaged metal films. As an example, the Ru film deposited on a ZnO single crystal is shown. Figure 6.7a shows AFM images of a Ru film of 40 nm thickness on a ZnO substrate. The RMS value of Ru that shows the clearest step-and-terrace structure, which follows the step-and-terrace structure of the ZnO substrate, is 0.14 nm. The pole figure image shows a Ru (hexagonal) phase (Fig. 6.7b). The peaks with six-fold symmetry at 60° in the ϕ scan can be seen at $\psi = 61°$ for the $(10\bar{1}1)$ plane of Ru, implying that Ru with a hexagonal structure grew epitaxially on the ZnO substrate with a c-axis normal to the substrate surface. A Ru composition spread film grew on ZnO epitaxially and showed the same crystal structure changes as the bulk phase diagram. By using FIBS, high-quality metal thin films can be obtained.

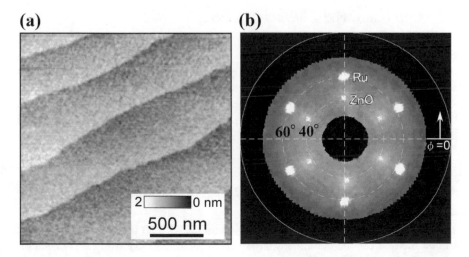

Fig. 6.7 AFM image and X-ray diffraction pole figure image of Ru thin film on ZnO substrate

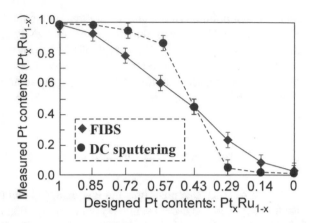

6.3.3   Combinatorial Thin-Film Synthesis by FIBS

By using FIBS with the combinatorial system (moving mask, target exchange system, substrate rotation, and PC control), a ternary alloy thin-film sample can be obtained, as shown in Fig. 6.3c. Furthermore, FIBS shows better composition spread linearity than the film using the conventional sputtering system, as shown in Fig. 6.8. Figure 6.8 shows the results of a composition analysis of the composition spread Pt–Ru binary metal films by energy-dispersive X-ray (EDX) analysis. The solid squares show Pt content of the film deposited by FIBS. These change linearly. Open circles show a referential result of a composition spread Pt-based metal film using the conventional sputtering system plus a combinatorial system.

6.4   Combinatorial Characterization

For the combinatorial synthesis technique, characterization is also important. Owing to the progress in measurement methods, by combining them with the PC-controlled stage system, we can apply several characterizations to the combinatorial research. Actually, in this decade, many measurement techniques have been applied to combinatorial thin-film syntheses. In particular, the two-dimensional X-ray diffraction method and AFM-based measurement methods are typical.

6.4.1   Two-Dimensional X-Ray Diffraction Method

The two-dimensional X-ray diffraction method (2D-XRD) is a suitable characterization method for the combinatorial synthesis technique. By using a 2D detector system, a part of the Debye Sherrer ring is two-dimensionally detected. In this system,

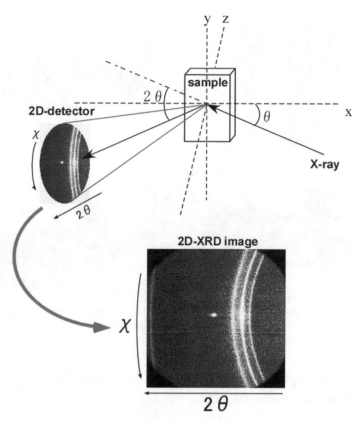

Fig. 6.9 Setup for 2D-XRD measurement and typical 2D-XRD image

the 2θ and ψ angles can be simultaneously detected. Figure 6.9 shows the measurement setup and a typical image of 2D-XRD. In the image, spot and ring patterns indicate a single-crystal structure including a twin structure and a polycrystalline structure, respectively; the typical measurement time for a 2D-XRD image is a few minutes. Thus, by combining a sample mapping stage, the 2D-XRD system can give us the mapping structural properties of the ternary composition spread metal thin films. For example, partial XRD data from a Pt–Ru–Nb ternary composition spread sample are shown in Fig. 6.10. The measured specific binary composition spread regions are indicated. The XRD intensity-2θ-composition plots are also shown for the Pt–Ru, Ru–Nb, and Nb–Pt lines in Fig. 6.10, which indicates that pure Pt was in (111) orientation, whereas Ru and Nd were polycrystals. Alloys of Pt–Ru and Pt–Nb also grew in a polycrystalline manner.

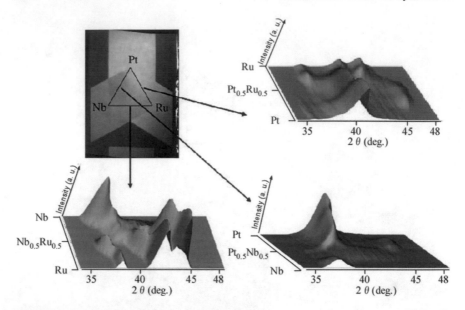

Fig. 6.10 XRD mapping results (sample photo and line scans on the Pt–Ru line, Ru–Nb line, and Nb–Pt line) of a ternary (Pt–Ru–Nb) composition-spread film (thickness 60 nm) on a SiO₂ (3 nm)/Si(1 0 0) substrate

6.4.2 Atomic Force Microscopy-Based Electrical Property Mapping Method

The atomic force microscope (AFM) is commonly used for imaging various surfaces of materials, such as ceramics, polymers, and even biological materials. The AFM consists of a cantilever with a probe tip used to scan a sample surface. When the tip is close to the sample surface, forces between the tip and the sample lead to a deflection of the cantilever according to Hooke's law [19]. Typically, the deflection is measured using a laser spot reflected from the top surface of the cantilever onto an array of photodiodes. By using the probe tip and the sample as a top electrode and an electrical ground, respectively, the AFM system can be applied to an electrical property mapping system, such as scanning nonlinear dielectric microscopy (dielectric properties) [20] and conductive AFM (resistivity). For the investigation of a metal/oxide interface, Kelvin probe force microscopy (KFM) is beneficial. KFM measures the local contact potential difference between a conducting AFM tip and the sample, as shown in Fig. 6.11a, thereby mapping the work function or surface potential of the sample. KFM is also used to study the band offset and/or electrical properties of a heterointerface such as a metal/semiconductor, as shown in Fig. 6.11b. The Schottky barrier height (Φ_B) is given by the difference between the potentials of the metal and the oxide. By combining the AFM-based electrical property measurements with a source measurement unit, we can obtain an electrical property map.

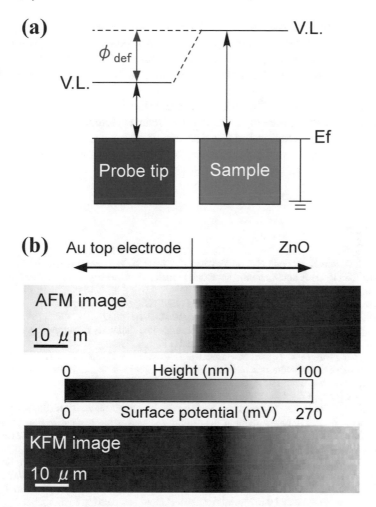

Fig. 6.11 a Schematic illustration of Kelvin probe force microscopy. The sample surface is grounded, which is equal to the Fermi level. Surface potential is estimated from the difference between the potentials of the probe tip and the sample (φ_{def}). **b** KFM image of Au/ZnO stack structure

6.5 Summary

The combinatorial thin-film synthesis is useful for high-throughput experiments of not only oxide thin films but also compound, metal thin-film materials. By combining the moving mask system, multitarget system, and PC control system, the combinatorial thin-film synthesis technique can be applied to the all physical vapor depositions. At the initial stage of development of combinatorial thin-film synthesis around 2000, the analysis technique limited the high-throughput experiments. Over the last decade, the analysis technique also progressed, achieving the high signal resolution and the small measurement spot area. Mass-produced systems also equip the moving stage and software visualizing measured results. By combining this analysis system and combinatorial thin-film synthesis, the new material development should be accelerated.

Currently, this technique assumes a large role in the material informatics. Material informatics consists of three topics. One is theoretical calculation. The second topic is the data mining including the big data and machine learning technology. The last one is the experimental approach. The combinatorial technique accelerates the systematic data store of materials.

References

1. Xiang XD, Sun X, Briceno G, Lou Y, Wang K, Chang H, Wallance-Freedman WG, Chen S, Schulz PG (1995) A combinatorial approach to materials discovery. Science 268:1738–1740. https://doi.org/10.1126/science.268.5218.1738
2. Schneemeyer LF, van Dover RB, Fleming RM (1999) High dielectric constant Hf-Sn-Ti-O thin films. Appl Phys Lett 75:1967. https://doi.org/10.1063/1.124887
3. Koinuma H, Takeuchi I (2004) Combinatorial solid-state chemistry of inorganic materials. Nat Mater 3:429–438. https://doi.org/10.1038/nmat1157
4. Chikyow T, Nagata T, Ahmet P, Hasegawa T, Kukuznyak D, Koinuma H (2010) Combinatorial oxide film synthesis and its application to new materials discovery. In: Oxide thin film technology-growth and applications, pp 37–57 (ISBN: 978-81-7895-468-4, Editor(s): Tomoyasu Inoue) (Transworld Research Network, 2010, India)
5. Cherief N, Givord D, Lie´nard A, Mackay K, McGrath OFK, Rebouillat JP, Robaut F, Souche Y (1993) Laser ablation deposition and magnetic characterization of metallic thin films based on rare earth and transition metals. J Magn Magn Mater 121:94–101. 10.1016/0304-8853(93)91157-3
6. Sakurai J, Hata S, Shimokohbe A (2005) Novel fabrication method of metallic glass thin films using carousel-type sputtering system. Proc SPIE 5650:260. https://doi.org/10.1117/12.581811
7. Ahmet P, Nagata T, Kukuruznyak D, Yagyu S, Wakayama Y, Yoshitake M, Chikyow T (2006) Composition spread metal thin film fabrication technique based on ion beam sputter deposition. Appl Surf Sci 252:2472–2476. 10.1016/j.apsusc.2005.05.078
8. Lippmaa M, Koida T, Minami H, Jin ZW, Kawasaki M, Koinuma H (2002) Design of compact pulsed laser deposition chambers for the growth of combinatorial oxide thin film libraries. Appl Surf Sci 189:205–209. https://doi.org/10.1016/S0169-4332(01)01002-9
9. Ahmet P, Yoo YZ, Hasegawa H, Koinuma T, Chikyow T (2004) Fabrication of three-component composition spread thin film with controlled composition and thickness. Appl Phys A Mater Sci Process 79:837–839. https://doi.org/10.1007/s00339-004-2627-9
10. Chikyow T, Ahamet P, Hasegawa K, Koinuma H (2003) Multi-element compound manufacturing apparatus. Japan patent, 2003-277914,A
11. Koinuma H, Matsumoto Y, Idaka K, Katayuama M (2006) Masking mechanism and film deposition apparatus having the same. Japan patent, 2006-063433,A
12. Yamamoto Y, Takahashi R, Matsumoto Y, Chikyow T, Koinuma H (2004) Mathematical design of linear action masks for binary and ternary composition spread film library. Appl Surf Sci 223:9–13. https://doi.org/10.1016/j.apsusc.2003.10.025
13. Mott-Smith H, Langmuir I (1926) The theory of collectors in gaseous discharges. Phys Rev 28:727–763. https://doi.org/10.1103/PhysRev.28.727

14. Heidenreich JE III, Paraszczak JR, Moisan M, Suave G (1987) Electrostatic probe analysis of microwave plasmas used for polymer etching. J Vac Sci Technol B 5:347. https://doi.org/10.1116/1.583900

15. Shatas AA, Hu YZ, Irene EA (1992) Langmuir probe and optical emission studies of Ar, O_2, and N_2 plasmas produced by an electron cyclotron resonance microwave source. J Vac Sci Technol A 10:3119. https://doi.org/10.1116/1.577874

16. Behrisch R (1981) Sputtering by Bombardment I. Springer, Berlin

17. Macak K, Kouznetsov V, Schneider J, Helmersson U, Petrov I (2000) Ionized sputter deposition using an extremely high plasma density pulsed magnetron discharge. J Vac Sci Technol A 18:1533. https://doi.org/10.1116/1.582380

18. Nagata T, Yoo YZ, Ahmet P, Chikyow T (2005) Effects of single-crystalline GaN target on GaN thin films in pulsed laser deposition process. Jpn J Appl Phys 44:7896–7900. https://doi.org/10.1143/JJAP.44.7896

19. Binnig G, Quate CF, Gerber C (1986) Atomic force microscope. Phys Rev Lett 56:930–933. https://doi.org/10.1103/PhysRevLett.56.930

20. Cho Y, Kazuta S, Matsuura K (1999) Scanning nonlinear dielectric microscopy with nanometer resolution. Appl Phys Lett 75:2833. https://doi.org/10.1063/1.125165

Chapter 7
General Summary

In this book, the research topics concerns the oxide thin-film materials for the electronics device applications, which were performed by author's grouped at NIMS. In each chapter, the following knowledge are confirmed:

Chapters 2 and 3: The interface structure of Schottky metal on oxide semiconductor.
The termination difference is effective on the metal/oxide formation. The oxygen-terminated surface can easily degrade the interface properties. Oxygen indicated the reduction process and/or passivation process is effective.
Chapters 4 and 5: The oxygen effect on the switching properties of ReRAM structure.
Oxygen plays a role of balance for the redox reaction at the metal/oxide interface. In the case of the nanoelectrite material combination, the density of the oxygen affects the ion migration in the materials. By changing the material combination, the density of the oxygen in the metal/oxide can be controlled.
Chapter 6: Brief introduction of combinatorial thin-film synthesis
In the former chapters, for systematic investigation, combinatorial thin-film synthesis was used. We briefly introduced the combinatorial synthesis and analysis. Especially, our developed combinatorial ion-beam deposition technique was mainly introduced. By using combinatorial technique, systematic investigation and acceleration of material development is achieved. This technique is suitable for the analysis of the hetero interface.

In addition, to investigate the interface structure nondestructively, hard X-ray photoelectron spectroscopy (HX-PES) was used. All HX PES experiments were performed at the SPring-8 BL15XU undulator beamline (NIMS beamline).

On the basis of above results, in the oxide-based electrical thin-film materials, oxygen plays an important role as well-known. Our results also imply to control the oxygen function. The surface termination, surface passivation, and material combination are effective in addition to the process conditions such as growth temperature and process ambient. This knowledge is important for nanoscale engineering of the electrical materials.

© National Institute for Materials Science, Japan 2020
T. Nagata, *Nanoscale Redox Reaction at Metal/Oxide Interface*, NIMS Monographs,
https://doi.org/10.1007/978-4-431-54850-8_7

Printed in the United States
By Bookmasters